D0464170

Using Mobile Technology to Deliver Library Services

A handbook

Using Mobile Technology to Deliver Library Services
A handbook

Andrew Walsh

THE SCARECROW PRESS, INC.
Lanham, Maryland • Toronto • Plymouth UK
2012

SCARECROW PRESS, INC.

Published in the United States of America
by Scarecrow Press, Inc.
A wholly owned subsidiary of
The Rowman & Littlfield Publishing Group, Inc.
4501 Forbes Boulevard, Suite 200, Lanham, Maryland 20706
www.scarecrowpress.com

Estover Road
Plymouth PL6 7PY
United Kingdom

ISBN 978-0-81088-757-2

First published in the United Kingdom by Facet Publishing, 2012.
This simultaneous US edition published by Scarecrow Press, Inc.

The paper used in this publication meets the minimum requirements of
American National Standard for Information Sciences—Permanence of
Paper for Printed Library Materials, ANSI Z39.48-1992.

Typeset from author's files in 10/13 pt Palatino Linotype and Myriad
Pro by Facet Publishing Production.
Manufactured in the United States of America.

Dedicated, of course, to Jenny and George,
my two lovely children.

Contents

Acknowledgements

First of all, I'd like to thank all the contributors of case studies for this book. Their individual and international experiences and insights make this work much richer than I could have done on my own. So thanks (in no particular order) to Anne Mary Inglehearn, Dina Koutsomichali, Lare Mischo, Graham McCarthy, Robin Canuel, Chad Crichton, Linda Barron, Sarah Pavey, Robin Ashford, Peter Godwin, Markus Wust, Tito Sierra, Mandy Callow and Kaye England for finding the time and energy to contribute their case studies.

I'm also grateful for the grant awarded to me by the John Campbell Trust, which part-funded a trip to the 2011 M-Libraries conference in Brisbane. Meeting and talking to many of the library staff at this conference who were using mobile technologies in their own work helped to inspire this book.

Last, but not least, thanks go to my employer, the University of Huddersfield, for encouraging innovation within Computing and Library Services, making it an interesting and challenging place to work.

Introduction and context

Introduction

The world of mobile technology has changed a great deal during the last few years, but mobile phones and mobile computing are not new. If we compare their uptake within libraries with, for example, uptake of internet use, it can be somewhat puzzling how little these technologies are utilized. Text messaging (SMS) was introduced on early mobile phones in 1993, at roughly the same time as browsers such as Mosaic enabled the emergence of the world wide web from the largely text-based pre-1990s internet. PDAs (Personal Digital Assistants) predated this, and Psion started to release these devices (which eventually merged with mobile telephony to create the smartphones that so many of us use today) from the mid-1980s onwards.

Libraries seem to have engaged enthusiastically with the early web, with many of us having webpages by 1995. Try looking on the Internet Archive Wayback Machine (www.archive.org/web/web.php) to find your library's early pages! However, most of us seem only recently to have started to think about engaging with mobile phones or mobile computing, despite the technology's having been around and available for a similar length of time. Now, increasing numbers of mobile devices are smartphones capable of accessing the internet and combining the capabilities of telephone and PDA. After this slow start, and as their capabilities grow, now is the time to give more serious consideration to taking advantage of these near-ubiquitous devices.

So, what can we do with these mobile devices that most – or all – of our users own? This chapter describes the context that we need to be aware of when considering how we can use mobile devices to deliver library services. With a clear idea as to how much this technology has become part of our everyday lives, we can then move on to the chapters that follow, which illustrate ways in which we can use the technology.

Context

Mobile phones now seem to be a near-ubiquitous technology. For example, there are more mobile phone contracts than people in the United Kingdom (130 contracts for every 100 people) and 92% of adults own and use one (Mintel, 2011a). Worldwide, in 2011 there were 5.9 billion mobile phone subscriptions and 79% of the population in the developing world owned mobile phones (International Telecommunication Union, 2011). To take just one of the most basic mobile phone functions, an incredible 4.1 billion SMS messages were sent daily during first half of 2009 (Van Grove, 2009).

The use of mobile devices is not limited to phones, of course, and the use of tablet computers, in particular, has exploded since the launch of the iPad. For example, almost 10% of UK adults owned a tablet computer in May 2011 (Mintel, 2011b), 19% of American adults owned one in December 2011 (Pew Internet and American Life Project, 2012a) and Gartner is forecasting that tablet computer sales will be in the region of 103.5 million units in 2012 (as reported in the *Guardian* by Arthur, 2011). Add to that the number of handheld games devices, e-book readers (19% of US adults owned an e-book reader in January 2012 (Pew Internet and American Life Project, 2012b)), netbook computers and more that we could include as mobile learning equipment, and the range and number of such devices is incredible.

These devices are driving a shift towards mobile internet access. Mobile data traffic in 2011 was eight times the size of the global internet in 2000 and, according to forecasts, mobile devices will soon outnumber human beings (Cisco, 2012). Active mobile broadband subscriptions reached almost 1.2 billion towards the middle of 2011, with mobile internet access capable of reaching 90% of the world's population (International Telecommunication Union, 2011). Mobile phone and internet access is allowing many developing countries to leapfrog the problems of installing fixed phone lines and to access the internet for the first time, thus revolutionizing information access for many people.

So, the availability and ownership of mobile devices are now enormous. The vast majority of the world's population is within reach of mobile phone and mobile internet connectivity. Most of our users have at least one device such as a mobile phone, netbook, e-book reader, handheld games device or portable music player that can be used as a mobile learning device.

Libraries have until recently considered these devices more as a source of noise and irritation than as vehicles through which to deliver their services.

'No mobiles' signs still proliferate in libraries, though they are slowly changing from outright bans to warnings to switch devices to silent operation. Now that mobile phones are an accepted part of everyday life for most library users, with increasing numbers of users owning tablet computers and e-book readers, we need to consider what this means for their potential as learning devices and for delivering library services. These devices are simply too much a part of our users' lives for us to try to ignore or even ban them, when we are trying to provide relevant and accessible library services.

There are two key streams of literature and thought about mobile learning, as there doesn't seem to be an overall consensus on what it actually is. There is one stream that tends to discuss it in relation to the technology. For example, one definition, by Traxler (2005), states that 'Mobile learning is any education provision where the sole or dominant technologies are handheld or palmtop devices'. The alternative stream gives greater consideration to the flexibility of the learner's situation, rather than focusing primarily on the technology – for example, the definition by O'Malley et al. (2003): 'Mobile learning is any sort of learning that happens when the learner is not at a fixed, predetermined location, or learning that happens when the learner takes advantage of the learning opportunities offered by mobile technologies.' This book tends towards the second definition, looking at the technologies that are available to facilitate learning and the delivery of library services. At times the technology enables new services, but often it simply creates opportunities for our users to access existing services in new and exciting ways, wherever they may be.

There is a great deal of literature available on mobile learning in general, with somewhat fewer readings available on mobile technologies' potential to deliver mobile services. Some general materials on mobile learning, and on libraries in particular, are included in the 'Further reading' part of this chapter. Much of the available literature is directly transferable to the library and information-provision environments and we should take full advantage of the available research. This is especially true of the literature that considers the learner as a whole, and mobile technology as simply a facilitator to enable learning to happen at any time and any place. This book takes elements of all these ideas and puts them into a practical and relevant context to help you to develop mobile-friendly services within your own library. At the moment much of the library-based literature focuses on the technologies. Many existing case studies cover the details of how a particular technology has been introduced. Relatively few discuss how access to our library services has been enabled by using a range of suitable technologies and why they were chosen to meet the users' needs. The focus has been on the how rather than on the why, or even on the user. The wide range of examples in this book, together with an emphasis on addressing users' needs, offers more balance between the technology and its suitability than previous texts tended to. The book extends individual case studies beyond consideration of old services forced to fit mobile technologies, to cover more of the context of the wider service and user needs.

Outline of this book

In the context of the widespread ownership and use of mobile devices; the growth of accessibility to mobile internet networks and Wi-Fi; and a history of educationalists taking advantage of these opportunities to enable mobile learning, the remaining chapters in this book look at how these opportunities can also apply to library services.

Written from the personal experience of the author – an academic librarian and teaching fellow at a medium-sized UK university – many of the examples are drawn from the context of college and university libraries. The services discussed, the examples and the case studies cover a wide range of UK and international libraries, from school settings to state libraries. It is hoped that in whatever sector of the overall library community you work, wherever in the world, you will find useful tips, examples and information to help you to improve your mobile offering. The material in the book is non-technical and should be accessible to anyone with an interest in the topic.

Chapter 1 looks in depth at something that is vital to the success of any new library service: considering your users' wants and needs. 'What mobile services do students want?' is based on research done at the University of Huddersfield, updated and expanded to be relevant to any sector. In any project, especially one to do with new technology, we shouldn't get carried away by our own enthusiasms. Instead, we need to consider what services our users want and would find useful, and focus on those. Everyone should have these thoughts foremost in their mind before introducing any new mobile project.

Chapter 2, on mobile information literacy, addresses an issue of real concern. When I was at school I used only print materials, either hand-outs from the teachers or the occasional book from the public library. As an undergraduate at university in the late 1980s and early 1990s I started using the internet for the first time, but again, my studying was from print materials (almost entirely books) from the library. Academic research and study was something done in the library, or in my room using materials on loan from the library. Finding out about other things, such as what was on at the cinema or what time a venue opened, involved either visiting the location itself or finding a printed phone book, going to a public telephone box and ringing someone up.

This chapter discusses the development of information literacy in the new, mobile-dominated environment. It is a first step towards considering what has changed between 'fixed' information literacy, where we seek information in set locations, and truly mobile information literacy, reflected in the truly mobile world. It is based primarily on existing literature and, as such, focuses largely on the search process itself, reflecting on how our users may be searching for information via mobile devices. If we can begin to realize what it means to be information literate now, rather than what information literacy looked like 20 years ago, we can begin to consider how this impacts on our provision of library services.

We move on from there to Chapter 3, 'The mobile librarian'. This discusses how librarians and information specialists can become more mobile. Covering areas such as mobile reference and roving, mobile productivity and keeping up to date, it gives practical examples for librarians who wish to take advantage of mobile opportunities in their own work.

If any of us were asked to name a mobile device, there is a high chance that we would name the mobile phone. If asked to name something we did on our mobile phone, many of us would quickly identify text messaging. Chapter 4 provides some ideas for how we can use this basic, accessible and well-used technology in many different ways.

Chapter 5, 'Apps vs mobile websites', discusses the issues surrounding the provision of information and services online to mobile devices. It outlines the advantages and disadvantages of two competing ideas – whether we should provide mobile-friendly websites that can be accessed from any device, or whether we should produce device-specific apps. It includes examples from several libraries to help you to decide which is best for your library service.

QR codes and location-aware applications have really taken off, with the advent of smartphones. QR codes appear to be everywhere at the moment in the UK, and libraries around the world have used them to market or provide access to their services. Increasing numbers of apps for Apple and Android devices seem to want to access the GPS chips embedded in these devices in order to calculate your location, and Google even tweaks your search results so as to take location into account. Chapter 6, 'Linking physical and virtual worlds via mobile devices', considers QR codes, RFID, location-aware services and Augmented Reality to show some potential and practical applications of these technologies. It discusses some quick, free and easy ways to use QR codes. For those libraries that use RFID tags to control their stock, it suggests ways in which to use this technology to provide alternative services. The chapter also outlines ways in which we should be considering location-aware services and how we could be preparing for the Augmented Reality services that are now emerging.

Chapter 7, on 'Mobiles in teaching', introduces some of the multitude of ways in which you could use mobile devices in the classroom, in particular to teach information skills. Including many practical examples and case studies from creative teacher-librarians, this chapter aims to equip every librarian who teaches with examples that they can apply in their own practice. The examples and ideas given are generally either free or low cost, especially when they make use of our learners' own devices.

Continuing the theme of examples and practical illustration of services is Chapter 8, 'E-books for mobiles'. This is an area that is undergoing rapid change at the moment, with increasing numbers of books being sold for reading on e-book readers and tablets. The chapter discusses some of the issues surrounding the provision of e-books in formats that can be read on mobile devices (as opposed to on a standard computer) and some of the services that are available for libraries.

The book is rounded off with a concluding chapter entitled 'So what now?' After covering a range of possible library services that can be delivered to mobile devices and giving many examples of how this might be done, this final chapter invites you to consider your next steps. As the first chapters discuss, things are different in the mobile world, but, first and foremost, you should try to deliver services in the way that your users want and need. This chapter helps you to reflect and consider what your users want and what your staff can deliver. It advises you to start steadily, but to introduce full services rather than pilots. It reminds us that no service has to be a final, finished product, and that, for new mobile services, we should review and assess as we go. Finally, it encourages us all to keep an eye to the future.

At the end of each chapter there is a brief, annotated bibliography of some relevant useful resources, primarily journal articles. This introductory chapter is no different, except that the 'Further reading' lists mainly general texts that cover many of the issues discussed elsewhere in the book.

It is hoped that all who read the following chapters will gain new ideas on how library services can be delivered via mobile devices and how we can take best advantage of them to support both ourselves as librarians and our users. As mobile devices continue to penetrate into our daily lives, it won't be long before 'mobile' becomes an integral part of every library's offering. Whether you are taking your first steps towards this or are already a long way down that road, I hope that you enjoy taking this exciting journey with your staff and users.

References

Arthur, C. (2011) iPad to dominate tablet sales until 2015 as growth explodes, says Gartner, *Guardian*, (22 September), www.guardian.co.uk/technology/2011/sep/22/tablet-forecast-gartner-ipad.

Cisco (2012) *Cisco visual networking index: global mobile data traffic forecast update, 2011–2016*, www.cisco.com/en/US/solutions/collateral/ns341/ns525/ns537/ns705/ns827/white_paper_c11–520862.html.

International Telecommunciations Union (2011) *The world in 2011: ICT facts and figures*, www.itu.int/ITU-D/ict/facts/2011/material/ICTFactsFigures2011.pdf.

Mintel (2011a) *Mobile phones and network providers – UK – January 2011*, London: Mintel.

Mintel (2011b) *Desktop, Laptop and Tablet Computers – UK – August 2011*, London: Mintel.

O'Malley, C., Vavoula, G., Glew, J., Taylor, J., Sharples, M. and Lefrere, P. (2003) *MOBIlearn WP4: Guidelines for learning/teaching/tutoring in a mobile environment*, www.mobilearn.org/download/results/guidelines.pdf.

Pew Internet and American Life Project (2012a) *Adult gadget ownership over time (2006–2012)*, www.pewinternet.org/Static-Pages/Trend-Data-Adults/Device-Ownership.aspx.

Pew Internet and American Life Project (2012b) *A Snapshot of E-reader and Tablet Owners*, www.pewinternet.org/Infographics/2012/A-Snapshot-of-Ereader-and-Tablet-Owners.aspx.

Traxler, J. (2005) Defining mobile learning. In: Isias, P., Borg, C., Kommers, P. and Bonanno, P. *Mobile Learning 2005*, Malta: International Association for Development of the Information Society Press.

Van Grove, J. (2009) *WOW: 4.1 billion SMS messages are sent daily, via mashable*, http://mashable.com/2009/10/07/ctia-wireless-survey/.

Further reading

Attewell, J. and Savill-Smith, C. (2004) *Learning with Mobile Devices: research and development*, London: Learning and Skills Development Agency. This book of papers came out of the mLearn conference, a major international mobile learning conference, in 2003. Although the papers are now a few years old, they cover a range of areas and are a varied and good introduction to mobile learning. There doesn't appear to be any consistency in publishing the mLearn conference proceedings, but those that have been published all appear to be openly available online.

Attewell, J., Savill-Smith, C. and Stead, G. (2006) *Mobile Learning in Practice: piloting a mobile learning teachers' toolkit in further education colleges*, London: Learning and Skills Development Agency.

Packed full of useful information and results of a major mobile learning project, this publication reports on the development of a mobile learning teachers' toolkit for post-16 education. The toolkit included a text-messaging quiz tool, a tool to create mobile learning games and the mediaBoard tool, which allowed a variety of media to be added to an online noticeboard and to material uploaded there by tutors. Besides the tools (which can rapidly become outdated in the fast-moving mobile environment), some really useful research questions are addressed on the topic of the impact of mobile learning on tutors and students. These results are still relevant today and the book is a highly recommended read for anyone introducing mobile services.

Griffey, J. (2010) *Mobile Technology and Libraries: the Tech Set #2*, New York: Neal-Schuman.

This fairly short book is part of the Tech Set series. It covers a range of issues to do with introducing mobile technologies into libraries, including planning, implementation, marketing, best practice and how to measure success, all from an easy-to-understand and basic perspective. It aims to be a primer for those considering the introduction of mobile technologies into their own libraries.

Hanson, C. (2011) *Libraries and the Mobile Web*, Library Technology Reports, American Library Association.

Part of a range of Library Technology reports, this one focuses on the issues that libraries should be aware of when introducing services that take advantage of the increasing accessibility of the mobile web.

Horrigan, J. (2009) *Wireless Internet Use*, Pew Internet and American Life Project, www.pewinternet.org/Reports/2009/12-Wireless-Internet-Use.aspx.

This project gathers a great deal of information on American use of the internet, including wireless and mobile access. If you want some quick and easy statistics from a highly reliable source in order to back up your arguments for a mobile library service, this is a good place to start.

Johnson, L., Adams, S. and Cummins, M. (2012) *The NMC Horizon Report: 2012 Higher Education Edition*, Austin, Texas: The New Media Consortium.

The New Media Consortium (NMC) produces a range of reports showing how it feels new technologies are travelling, and their uptake. The reports are all freely available to download from the NMC's website. This Higher Education 2012 edition outlines a range of technologies that NMC feels will be mainstream in higher education over three time periods: in one year or less; in two to three years; and in four to five years. Under 'one year or less' are both mobile apps and tablet computing, with mobile tools being part of many of the other future promising technologies. If you aren't already aware of the NMC reports, they are a great resource for helping to identify emerging technologies that are likely to have a major impact on us all.

Kroski, E. (2008) *On the Move with the Mobile Web: libraries and mobile technologies*, Library Technology Reports, American Library Association. Another Library Technology report, this one covers a whole range of technologies. Starting from scratch, Ellyssa Kroski outlines an enormous range of potential applications for mobile services, many of them fairly basic. Most of the references in this report are to web addresses, especially news stories and blogs. That said, the overview is broad and shows the sheer range of services that people were considering several years ago.

Library Success Wiki (2012) *M-libraries*, www.libsuccess.org/index.php?title=M-Libraries.
This wiki is a useful resource that links out to a range of examples, papers, conferences and more on mobile library subjects. Like many wikis, it can be out of date in parts and may be updated only intermittently, but there had been several edits to the page in the week that this list of references was prepared. If you find it useful and want to keep it up to date for other people, please contribute to the effort of doing so!

Mills, K. (2009) *M-Libraries: information use on the move: a report from the Arcadia Programme*, http://arcadiaproject.lib.cam.ac.uk/docs/M-Libraries_report.pdf.
A basic and very accessible general introduction to mobile library services, based on Keren Mills' Arcadia project. This was based on a survey of staff and students at two very different UK universities: Cambridge (very traditional) and the Open University (entirely distance learning).

Murphy, J. (2011) *Mobile Devices and the Library: handheld tech, handheld reference*, New York: Routledge.
This book, edited by Joe Murphy, who regularly presents on mobile technology and libraries, was originally published as a special issue of *The Reference Librarian*. It is a selection of papers from the Handheld Librarian conference. There are other special issues of *The Reference Librarian* containing the papers from other Handheld Librarian conferences. This series of conferences is online and attracts an international (though largely American) audience. If the M-Libraries conference is the key annual conference in this area, I'd suggest that the Handheld Librarian conference is a much cheaper and more accessible alternative.

Naismith, L., Lonsdale, P., Vavoula, G. and Sharples, M. (2004) *Literature Review in Mobile Technologies and Learning: report 11. Educational technology,* Futurelab, www.futurelab.org.uk/resources/documents/lit_reviews/Mobile_Review.pdf.

In the further reading at the ends of the other chapters, you should see that the majority of references are recent ones. I've tried to show what is currently happening as exemplars. In this reference, along with some of the others in this section, we can see that there is a longer history to mobile libraries and mobile learning. This report is a high-quality literature review on mobile learning, helping to set the subject in the context of earlier pedagogical research. This is an extremely useful report for those of us who want to think about mobile services in a wider pedagogical context.

Needham, G. and Ally, M. (2008) *M-Libraries: libraries on the move to provide virtual access,* London: Facet.

Ally, M. and Needham, G. (2010) *M-Libraries 2: a virtual library in everyone's pocket,* London: Facet.

Ally, M. and Needham, G. (2012) *M-Libraries 3: transforming libraries with mobile technology,* London: Facet.

These books are essentially the conference proceedings of the M-Libraries conferences. These international conferences attract an excellent mix of papers from around the world and from every library sector. The annual M-Libraries conferences themselves are the key international mobile library conference to attend each year. I've deliberately not included any of the individual papers from these in any of the following chapters. Instead I'd recommend obtaining the full books and yourself selecting those papers of most interest.

Walsh, Andrew (2009) They All Have Them – Why Not Use Them? Introducing mobile learning at the University of Huddersfield Library. *Sconul Focus,* **47** (Winter), 27–8.

Walsh, Andrew (2009) Text Messaging (SMS) and Libraries, *Library Hi Tech News,* **26** (8), 9–11.

Walsh, Andrew (2009) Quick Response Codes and Libraries, *Library Hi Tech News,* **26** (5/6), 7–9.

Walsh, Andrew (2010) Mobile Technologies in Libraries, *FUMSI* (34).

Walsh, Andrew (2010) QR Codes – Using Mobile Phones to Deliver Library Instruction and Help at the Point of Need, *Journal of Information Literacy,* **3** (1), 55–65.

Walsh, Andrew (2010) Supplementing Inductions with Text Messages, an SMS 'Tips and Tricks' Service, *ALISS Quarterly,* **5** (3), 23–5.

Walsh, Andrew (2010) Mobile Phone Services and UK Higher Education Students, What Do They Want from the Library? *Library and Information Research*, **34** (106), 22–36.

Walsh, Andrew (2011) Blurring the Boundaries between Our Physical and Electronic Libraries: location aware technologies; QR codes and RFID tags, *The Electronic Library*, **29** (4), 429–37.
A selection of articles on mobile technologies and libraries by the author of this book. Whenever copyright permissions permit, all my articles, book chapters, major conference papers and more are uploaded to the University of Huddersfield institutional repository at http://eprints.hud.ac.uk and are freely available for anyone to download.

1

What mobile services do students want?

Introduction

I carried out a study in late 2009 within my own library environment, looking at students' attitudes and their willingness for the library to intrude upon something that might be viewed as a very personal tool, that is, their mobile phones. Together with Chapter 2, on mobile information literacy, this chapter helps to put the ideas and case studies in this book into context. It gives a flavour of the sorts of services university students consider useful and important, many of which would transfer to other sectors. This is an area that is continuing to evolve as the penetration of high-level mobile devices such as smartphones and tablet computers increases.

Context

This was an important area to study because, although many mobile learning and libraries case studies are published in the literature, most of them focus on the implementation of a technology or service. Many projects have supplied mobile phones or PDAs (personal digital assistants) to trial participants, for instance the Assessment and Learning in Practice Settings (ALPS) project, which issued 900 high-end mobiles/PDAs to study participants.[1] There has thus been limited study of the level of these services' acceptance among students using their own mobile devices in the context of the delivery of library services, in particular of whether students would see contact initiated by the library via their own mobiles as intrusive, as opposed to services such as 'text a librarian', where the students themselves choose to initiate contact. There are small elements of this in some of the existing literature, for example Uday Bhaksar and Govindarajulu (2008) report some brief examples of student feedback on the use of text messaging (SMS) services and Pasanen (2002) describes the early adoption of such services at Helsinki University of Technology.

However, in the commercial sector there has been relevant research on contact by companies, particularly promotional contact, with their customers. Some of this can be directly translated into potential library uses. A Finnish study led by Merisavo (2007) found that mobile advertising that recipients perceived to be useful in regard to both context and content was generally well received. Merisavo also looked at issues of control and trust, that is, whether mobile owners feel some sense of intrusion and perhaps powerlessness on account of receiving advertisements from perhaps dubious senders. The study found that, as long as messages were perceived to be useful in relation to both information and situation/timing, neither control nor trust was a significant factor.

A further, rather more complex study led by Karjaluoto (2008) broadly concurs with some of Merisavo's findings, but also brings in 'perceived social utility' as an additional driver, or as an extension of the message's usefulness. Another slight difference in findings is that consumer trust, which Merisavo considered relatively unimportant, is believed to develop gradually as organization–consumer interactions increase, and so can 'solidify the relationship', thus fostering 'mutually beneficial exchange'. Also Karjaluoto, drawing upon his earlier 2006 study on demographics (see above), looks at intention in relation to gender, age, education, income and even household size, and again cites relative youth as increasing the likelihood of a positive predisposition towards mobile marketing. The results of these studies suggest that in libraries, which are normally perceived as a neutral, trustworthy space, it would naturally be more likely for text messages to be seen as acceptable, and that this acceptability would be reinforced in academic libraries, where the key demographics match those which, in a previous study (Karjaluoto et al., 2006), Karjaluoto found to be particularly comfortable with mobile contact – that is, people in the age ranges 16–20 and 21–25. The element of 'usefulness' is one that is important for any new service that a library might provide, and especially relevant for library services that make use of mobile devices.

The use of SMS 'reminders' is also creeping into education in general, with schools, colleges and universities experimenting with text messages to remind students about deadlines and more. A study by Jones et al. (2008) showed widespread acceptance of text message reminders amongst their students, and is directly relevant to one way in which libraries may choose to use mobile technologies.

In terms of the potential for libraries to send out messages longer than simple reminders or reservation notices, a piece of largely French-based research was conducted that investigated whether and when a saturation point develops beyond which mobile advertising is, at best, non-efficient, or possibly even a source of irritation to recipients (Gauzente et al., 2008). The study's cautionary findings were that a complex relationship exists between demographics, how often and for how long people used their mobiles, and the frequency and length of messages sent, all of which have a bearing on when this saturation point may be reached. Again, though, concurring with the above studies by Karjaluoto and Merisavo, the study found that usefulness is the most important factor in the adoption of SMS.

Details of the study

As the mobile phone is so personal to many people, I thought that it would be preferable to take a qualitative approach to gathering much of the data and to give the students a chance to discuss and present their feelings on this topic in their own words.

The primary research method used was focus groups, a good way of exploring feelings and expectations with the population studied, as members are able to interact with each other and develop ideas that might not be expressed individually. The limitations of this method include the risk of one or two members dominating a group, or of people being reluctant to express opinions contradictory of those already given (so that a 'group-think' mentality can arise). However, our library service uses focus groups regularly and has a long track record of gathering information through such groups in order to improve its services, along with corresponding experience of moderating these groups in an effort to reduce some of the key limitations.

Students were recruited across a range of courses based full or part time at the main university campus, both undergraduate and taught postgraduate. They were invited to small focus groups. A prize draw for an iPod Nano provided a small incentive to attend and refreshments were provided during each meeting. Recruitment was through advertising the focus groups and associated prize draw via the Library's Twitter account, subject team blogs, plasma screens within the Library and student library inductions across all subject areas. Recruiting students was problematic, as was turning an expressed willingness to attend a focus group into actual attendance. In total, 18 students attended the groups.

Five focus groups were held in the autumn term of the 2009–10 academic year. Initially we focused on the idea of Library contact via SMS to the students' own mobiles and the groups were asked to discuss the idea and to think about the issues that might arise with this sort of contact. They were then presented with some possible services, including but not limited to text messaging, along with brief explanations and examples where possible. They were asked to discuss whether they felt each service would be interesting and useful to them and their peers. Finally, the focus groups were asked to rank a list of ten potential, mobile-friendly services in order of priority for development, with the service they felt as a group to be most useful ranked at number one, and the service they considered to be of least utility ranked at number ten. They were also invited at this stage to suggest other potential developments for the Library to consider.

The comments and concerns expressed in the focus groups were grouped and analysed in order to bring out the key concerns and attitudes with regard to Library contact via SMS.

One additional piece of data gathering was carried out alongside this study, a one-day exit survey of the Library. This addressed directly the question 'Would students accept contact via SMS as a "default" option, or would they prefer "opt-in" services?' as it was felt initially that this question lent itself to a quantitative approach by which we could easily gather a large amount of data in response to a simple 'either/or' question. The researcher stood in the area immediately beyond the Library exit and exiting users were asked whether they'd be happy for the Library to contact them by SMS using the details on their student accounts, or would prefer to have the choice of opting in to such a service.

Attitudes towards text messaging

Attitudes towards text messaging from the Library were overwhelmingly positive. There were some concerns with the Library's using SMS, but these were based on whether the messages would be 'useful' or promotional. Only one group brought up the issue of the text messages' being potentially intrusive, making comments such as 'I prefer, I mean, text messages for me are quite personal, they are from friends not institutions ...' and 'I get annoyed if I get a text message from my network, do you know what I mean? I pull out my phone, see what it is and think, was it really necessary for O2 to send this right now?'

Even the group that raised concerns about the potential intrusiveness of text messages stated that they'd be happy to receive text messages by default, as long as the messages were useful, and ranked text messaging services at third and fourth in desirability, out of 10 suggested mobile services. This concept of 'usefulness' cropped up again and again in the focus groups, with all groups considering it acceptable for their university to introduce SMS-based services for all students (so offering only an opt-out option, not waiting for users to opt-in), as long as the service was perceived to be 'useful'.

Perceptions of 'usefulness' varied slightly between the groups. Some concrete examples were:

'Reminders and things like that would be quite useful.'

'I personally wouldn't mind receiving notifications and things, because I'd find it useful.'

'I received a message from the University reminding me I was working tomorrow and I found that really useful.' (From a student employed by the University on a part-time basis.)

'If you booked a room and were being told it was free, that would be okay, but if you were texted by IT to say something's down and that happened frequently then you'd get a bit annoyed.'

'If you requested a book and it's come in, it saves you from having to, if you don't have access on your phone, it saves you having to find a computer to find out if your book's come in. If you get a text, you'll know you've got it and if you're out and about you can just pop into the Library and pick that book up.'

'That's stuff you're actively interacting with the Library with, so you've requested a book or booked a room rather than the Library cold calling you on your time.'

All of the groups agreed that the library services that they had already chosen to interact with were the ones for which they'd most like to see SMS contact. These services included loans (i.e. notices about books being due back, or overdue); notices about their requested items becoming available (so that they knew to come into the library and pick them up); and messages about room bookings (to remind them that their booking was about to start). In general, these are services currently dealt with by e-mail reminders, and it might be that common library e-mail reminders could fairly readily be duplicated by text message, as libraries can be fairly confident that if the e-mail notices are perceived to be useful, then the text messages are likely to be well received.

At my request, the issue of whether to target all students or only those who chose to opt in was discussed at length. More concerns were expressed about students potentially missing out than about the use of students' mobile numbers by the University. As long as the service was perceived to be useful, all the groups felt strongly that it should be introduced for all students, by default. Concerns were repeatedly raised that students might miss out if they had to choose for themselves whether or not to subscribe to a text messaging service:

> 'I can see the people that are more likely to forget their library books are those that are also most likely not to opt in.'

> 'I think that if you do it, it has to be driven by the Library itself. If you offer the option to students, then 9 times out of 10 they won't, either through forgetting or just not wanting to do it.'

> 'I think it's important that you guys drive it ...'

Besides the issue of 'usefulness', the only other reservation expressed about all students automatically receiving text messages using the numbers harvested from their student records was that it should be easy to opt out if required. The opt-out should also be clear and easy to do; it should be made clear to all students that they would receive text messages, the reason for this should be explained and there should be instructions on how to stop receiving the messages:

> '… if you could just be like other services and you could text back "stop", then that would be okay.'

At first glance, the exit survey somewhat contradicted the focus groups, with only 46% (n=150) of users stating that they'd be happy for the Library to use existing records to automatically contact all students, and 52% saying that they'd prefer to have to opt in. This reflected the comments made in the focus groups about any contact having to be seen to be useful to library users. It is therefore believed that it was too simplistic for the exit survey to ask a yes/no question with a response tightly tied to whether students could immediately see text messages being of utility to them, and without their being given the chance to consider examples.

General feelings about mobile services and the Library

Some general feelings about using the Library and mobile phones came out of the focus group discussions that were perhaps a little unexpected. They were not initiated by the researcher, but emerged from the general discussions:

1 **A perception that you only interact with the Library while in the Library.** Mobile services are often introduced in order to help busy students who are thought of as being constantly on the move and

needing opportunities to interact with services wherever they are, in whatever small amounts of time they have available. We expect that they may want to use mobile library services on public transport on their way to lectures, in snatched moments between lectures or while waiting to be served in a supermarket queue. In this research, however, the comments from the participants frequently implied that they thought the only place where people would be interested in interacting with the Library was from within the Library. On 'text a librarian' services, comments were made by several participants such as '… didn't see the point. Obviously there are librarians knocking about all over the place', and 'if I go to the library and need to ask someone something …'. These ideas cropped up during discussion of several potential services. One participant commented on a potential text-message tips service: 'personally I'd prefer like an FAQ thing on a piece of paper', and another said, 'if you had it in leaflet form, in an obvious place, like in the middle of the room …'. This perception seemed to persist to some extent across all the groups, and there was a rather unexpected attitude that, although we were discussing mobile services, users would primarily use them only when they chose to come in to the Library. This may mean that we should be cautious in providing some mobile services. If potential users of a mobile service wouldn't really consider using it outside the library, this has serious implications. Perhaps when planning mobile-friendly services we should expect them to be used mainly within the library.

2 **A reluctance to use the mobile web.** More than half (55%) of focus group participants had accessed e-mail or the general mobile web on their phones, but they seemed reluctant to use mobile internet access unless they had a concrete reason for doing so. It was seen as difficult and potentially costly by several participants, and comments were voiced such as 'depends who's paying for it' and 'if it's through your phone it can cost a fortune'. It may be that even as mobile internet access becomes easier and more widespread and costs come down (charges are already normally capped at a fairly low cost per day in the UK, even on pay-as-you-go phones), the perception of mobile internet access as costly and difficult may persist for some time. The University has wireless internet that can be accessed through most smartphones, which would have made internet access free for the smartphone users in the groups.

3 **Enthusiasm about using potential services that would meet a perceived need – immediately contradicting points 1 and 2 above.** One potential service was mobile search of library services, using a mobile version of Summon (the Library's search tool for electronic resources). Many of the participants were quite enthusiastic about this

possible service, suggesting uses for it outside the Library via their own mobile phones (despite having previously expressed reluctance with regard to accessing the mobile web) or for accessing mobile-friendly Library services in general outside the Library building. This reflected a common problem when developing new services. Often potential users cannot imagine new services when asked, but when presented with services that could be useful they will enthusiastically adopt them.

4 **Little sense of a desire to explore or experiment with new services.** Some possible services, such as QR codes, were described as potentially being useful, even 'futuristic' (despite their already being available for several years), but the groups showed no desire to experiment with or explore these services unless they were already convinced of the services' usefulness to them. This is likely to provide a significant barrier to the introduction of any new service if we cannot rely on our users to try new services without first persuading them that they will find those services useful. The very nature of new services means that it is hard to do this: we really want users to find out for themselves, allowing us to then fine-tune services and persuade others of their usefulness, based on current users' response.

Potential services

A range of potential mobile phone services, prepared by me, was discussed and demonstrated (where possible) and each focus group discussed its feelings about these potential services and was asked about any other services it would like to see developed. The groups were also asked to rank these potential services in order of priority for development, with number one being the most desirable. The rankings of each focus group were combined by treating the ranks from each group as scores and adding them together. The potential service with the lowest score was then ranked as number one (most desirable). The overall, combined ranking for the potential services was as follows:

1 Reminders by text (e.g. for overdue items)
2 Search from a mobile device (to easily search the catalogue or electronic resources)
3 Renewals by text (text a central number to automatically renew all items borrowed)
4 More mobile-friendly web pages (to be more accessible from mobile devices)

5 Help by text message (a 'text a librarian' service)
6 Tips by text message (a series of text messages to support inductions or
   information skills)
7 Vodcasts (video materials, but in mobile-friendly formats)
8 Podcasts (audio materials easily downloadable to mobile devices)
9 QR codes (codes that can be read using freely available applications on
   mobile phones and used to link to further information: web resources
   and contact details, amongst other things)
10 Bluetooth (to automatically recognize users as they walk through the
   library and deliver appropriate materials or alerts to their mobile
   phone).

There was one additional suggestion from one group: to be able to search one
shelf at a time from a mobile device within the Library that is, to restrict a
catalogue search to particular physical locations. The group ranked this as its
second choice.

Table 1.1 shows a break-down of the rankings by focus group.

Reminders by text

This service was by far the most popular, with three out of the five groups

Table 1.1 *Potential mobile library services, ranked by students in order of preference*

	Group A	Group B	Group C	Group D	Group E	Overall RANKING
Reminders by text	1	4	1	3	1	1
Search from a mobile device	3	3	6	2	2	2
Renewals by text	2	7	2	4	3	3
More mobile-friendly web pages	8	1	5	1	4	4
Help by text message	7	5	3	7	5	5
Tips by text message	6	6	4	8	7	6
Vodcasts	4	10	7	5	6	7
Podcasts	5	9	8	7	8	8
QR codes	9	5	10	6	10	9
Bluetooth	10	11	9	10	9	10
Searching by shelf	x	2	x	x	x	

making it their first choice. This was the service that was viewed in all the group discussions as having concrete benefits, and the one that most closely matched the sort of service the participants would like every student to receive automatically unless they decided to opt out. All groups saw it as being convenient and valuable to all students who used the Library. Suggested uses for reminders included when borrowed items were due back, when a requested item was available to be collected and when a group room that the user had booked became available.

Search from a mobile device

This potential service caused some excitement in the groups, with some unexpected suggestions as to how they thought it would be of use. Although there was a general feeling within the groups of reluctance to connect to the internet via mobile phone, there was an immediate feeling in most groups that they would make an exception for this. The researcher expected that this sort of mobile search might be used when library users were travelling or otherwise had small amounts of time to spare. Some participants reinforced this expectation:

'I commute a lot and print off a lot of journal articles to read on the train. If I could read them on my phone, that would be really useful – I wouldn't have to print it off and pay for it, as well as it being more accessible.'

'You wouldn't have to come all the way here just to see if there was a book in.'

However, an unexpected suggestion that came from two groups independently of each other was its use in lectures to look up references immediately they were mentioned by the lecturer:

'I'd probably use that quite often in lectures. When a lecturer recommended a book I'd pull out my phone and maybe add it to a list or something to use it later on.'

'If you were sitting in a lecture and wanted to see what journals were available later, then it would be useful.'

Another, somewhat unexpected suggestion, was its use within the Library to save walking the (rather short) distance to access a fixed library catalogue, with several groups stating that it would be incredibly convenient to have while browsing the shelves or studying within the Library:

'Sometimes, when I'm in the Library, if I'm sitting down in the music section, then the library catalogue computers are near, like, the stairs, so to check if there's a book in, like, I'd have to go all the way there, but if I could access it where I am then that would make things a lot easier.'

Renewals by text

This was seen as valuable by all groups, and often linked with 'reminders by text' during their discussions. It was narrowly beaten in the group rankings by 'search from a mobile device', as there was some suggestion within the groups that it is so easy to renew books already (the Library allows renewals via the catalogue, by phone (voice), using self-service machines or in person), that it didn't provide as much benefit to users as would mobile search.

More mobile-friendly web pages

Quite realistically, this was discussed at length in various groups as being a necessary precursor to other mobile-friendly services. If the Library web pages weren't mobile friendly, then it was considered pointless to produce other mobile-friendly resources because it would be too hard to navigate to them:

'They have to be more mobile friendly before you can do the others.'

This seemed to be the key reason for ranking this service so highly – not because there was any desire to view the normal Library web pages in mobile-friendly formats, but because it was a necessary step for users to be able to find and access some of the other services. In general, there did not seem to be a desire to access the Library pages in a mobile-friendly format; rather, there was a desire to access certain *services* in a mobile-friendly way. The focus groups wanted the mobile web pages to link them easily to these services. Mobile-friendly web pages were regarded as preferable to dedicated apps, probably due to the variety of mobile phones used by the participants.

Help by text messages

While the other potential services described above were generally viewed positively and often discussed enthusiastically, this service (fifth in the ranking) and others below it were often seen as lacking relevance to library users. While it was generally seen as a good, convenient service, many participants couldn't see why it would be useful. Some illustrative comments have already been given above, showing that many students thought of themselves as only wanting to ask questions about the Library or to access information whilst in the Library. Other comments included:

> 'If I go into the Library and need to ask someone something it's usually more complex than can be answered in a simple text.'

There was also a feeling that speed was of the essence when answering text questions, and time scales of up to half a day were suggested as being acceptable:

> '... that's crucial, isn't it? I think two hours is unacceptable, more like five minutes ...'

> '... if you texted in the morning and had a reply by midday or whatever, then that would be fast enough to be useful.'

In general, there was an acceptance that a text message sent to ask the Library a question was not the same as texting a friend, and so most participants did not expect a response within minutes. A fairly quick response was expected, however.

All the students in the focus groups came to the campus regularly, so the lukewarm response to text-a-librarian types of service may be a reflection that they were more likely to ask for help face to face than to use other enquiry methods.

A little while after this study, we introduced several variations on the standard text-a-librarian service. We made it easier to text directly to services such as disability support or IT help. The service that has become most popular is our text 'NOISE' service. Library users can text-message to us the location of unacceptably noisy or inappropriate behaviour and we then send a member of staff to investigate. This links with other comments made by the focus groups. Users can immediately see a reason for using the text 'NOISE' service to report bad behaviour. As for the more general text-a-librarian (or IT, disability support etc.) types of service, they struggled to think of questions to ask via these services, as opposed to via our many other service routes.

Tips by text message

This 'drip feeding' of information by text was generally seen as a service that could be useful:

'That's definitely going to work because a little bit of information over a long period is far more likely to sink in.'

'There are some people on my course that have hardly been in the Library. If you do make them realize how useful it can be pretty early on, then you're going to put them in a better position, because right at the beginning is when they are likely to try new things, when they've just arrived.'

'... for first years, a really good idea.'

This was tempered by several people in the groups thinking about only wanting information on the Library while in the Library. Hence comments such as those below, essentially asking for a service that we already offer via short library handbooks displayed across the Library:

'If you had it in leaflet form, in an obvious place, like in the middle of the room, you could have sections saying what to do ...'

'I'd prefer, like, an FAQ thing on a piece of paper.'

These mixed views meant that some groups struggled to come to a consensus as to the relevance or priority of this service, and it tended to hover around the middle of the list of priorities. Any introduction of 'library tips' by text message would have to tread a fine line between being useful to the majority of potential users and too simplistic for some. Perhaps this would only truly succeed if targeted at particular groups of users. For example, in a higher education institution international students could be targeted during their first few months with a very tailored set of tips and this might help to make them more obviously relevant to potential users.

Vodcasts and podcasts

There were some mixed views on these possible services, including some limited concerns about connection charges:

'... coming through a wireless hotspot it's fine, but if it's through your phone it can cost a fortune.'

None of the groups could see why they would want to view video or listen to audio from the Library, though there was a slight preference towards video, particularly for showing more complex or confusing areas such as research techniques and search skills – potentially as a replacement for library information skills sessions. Across all groups, there was a definite preference for live streaming of content over downloading of material, with comments including:

> 'I don't know if a lot of people would go through the process of downloading it, they'd want a quick live thing and having to be transferred onto an iPod or whatever, would end up being more hassle than it's actually worth. I think YouTube live streaming is probably a better way of transfer.'

> 'I've streamed a couple of the videos on my laptop, everyone's used to going on YouTube and everything and videos playing, but I probably wouldn't download one.'

The issue of being interested in interacting with the Library only when in the Library was also raised again, and one group concluded that there was no point in watching such materials on their mobile devices because there were 'plenty of computers in the Library' and they'd prefer to watch them on the larger computer screens.

On a positive note, one student said that 'if it worked and helped me, I'd use it and show it to all my friends', bringing us back again to the importance of perceived usefulness for all these services. It seems that if we can persuade students that any of these services would be of concrete use to them, then they will be happy to use them.

Since this study was carried out, we have seen more 'dual screening' in the Library, with users carrying out some tasks on a fixed computer and others on their mobile phone or tablet computer. This may be the area where mobile-friendly video and audio materials would be of the most advantage to users. If they can carry out an activity on a fixed screen while following the instructions using their mobile, the materials will benefit the users clearly and directly. This could have implications for the types of videos and podcasts that we develop. If we design them primarily to be viewed on a second screen we may see them become more popular with students than ever before.

QR codes

All the groups felt that these were too complicated and that the barriers to their use were too high for them to be of widespread use – even though the only barrier is to install a free application onto almost any camera phone. It seems that, unless they are convinced in advance of QR codes' utility, users will not install an application with which to try them.

Some illustrative comments included:

'Why bother?'

'I don't really think enough people can use them ... it's not really going to happen.'

'Personally I saw it and didn't know what to do with it.'

The only positive comments about QR codes were summed up in one short statement by a student: 'They have potential, but ...'. No groups felt that they were an accessible and useful enough service to be worth the Library's spending time on them.

This suggests that while QR codes are free to introduce and may be worth investigating, they may be of limited use until they become commonly used in other settings. Because they are so easy to produce they are now starting to appear more regularly in many more settings, including in libraries. As they become more commonplace in other settings, they become more useful within libraries. If your users have installed a QR reader to view a movie trailer, or to enter a competition from a popular consumer product, they then become more willing to try it out in the context of your library. We discuss this in more detail later, and describe many of the exciting ways in which QR codes can be used in Chapter 6, on linking physical and virtual worlds.

Bluetooth

Comments about using Bluetooth in any way were overwhelmingly negative, and no students in any group thought that it was an appropriate medium for interacting with the Library, mainly due to only ever turning Bluetooth on when they want to carry out a specific task, such as exchanging information with friends or between devices:

'Most of the people I know never have Bluetooth turned on.'

'I only turn it on if I want to exchange information ...'

The most succinct comment summing up the feeling about Bluetooth, agreed by all members of a focus group, was 'Oh, no, that is horrible!' suggesting that whatever service was offered using Bluetooth, it was unlikely to have a significant uptake.

Conclusion

There are always dangers in generalizing results from one small study at an individual institution. However, the conclusions and implications for practice listed below are drawn from qualitative data that should be readily transferable to other institutions where the context of mobile phone use is similar. This includes institutions across the UK, Europe, North America and Australasia. It is unlikely that the conclusions will be as relevant in areas where current mobile phone availability and use follows significantly different patterns, such as in most developing countries and in certain highly developed countries in Asia (such as Japan). The ordering of preference for individual services may well vary from institution to institution, but the general attitudes towards text messages and mobile web-friendly services is likely to remain relatively constant within the general cultural environment of mass mobile phone use.

The students who participated in the focus groups confirmed some of the commercial research into the acceptance of text messaging contact. They were overwhelmingly positive about receiving text message-based services from the Library, with the key caveat being that they must believe that the services are *useful* to them. This was confirmed in discussions about all the possible services, with perceived and obvious utility being the most important factor when deciding if they were interested in a service's being developed. There was no sign of participants being willing to experiment and explore new services so that they could discover for themselves what services would be useful.

Participants were also reluctant to use the mobile web, even when able to do so, although this should already be changing as accessing the web via mobile phones becomes increasingly mainstream. Indeed, it was a little surprising in this study to find that there was still reluctance to use the mobile web. It may be a hang-over from those early adopters who were 'stung' by high charges when they first used the services.

The results suggest that libraries that are considering increasing their services to mobile users should:

1 Initially introduce services that use text messaging, rather than more sophisticated services via the mobile web. These are a 'quick win' and popular with the majority of users. Services on the mobile web should

then follow after the quick win of text messaging services.
2 Concentrate on services for which potential users can immediately see
   the benefits, such as 'reminders' of overdue books, rather than services
   with less obvious or less mainstream benefits.
3 Make sure that any mobile-friendly services are marketed carefully,
   selecting the groups most likely to benefit from them and directly
   stressing those benefits to potential users in any promotional activities.

This study qualitatively examined students' attitudes to mobile-friendly
library services in general, rather than focusing on individual services being
piloted or introduced by a library. This is unusual in the published literature
and there is potential for more studies in this area, investigating what
students would like to see developed rather than whether new or existing
services are working. It would also be beneficial to investigate if participants'
current wariness of the mobile web in this study is significantly reducing as
the market penetration of smartphones increases, which is happening rapidly
in most countries. Any such studies would help to build a consensus on the
sorts of mobile services that libraries should be developing, based on the
attitudes and desires of their users rather than on the preferences of
institutions and their funders.

This focus on users' requirements, needs and desires is something that is
lacking in much development of mobile library services. I would encourage
any library service to take note of the many possibilities outlined elsewhere in
this book, but the starting-point should always be to consider your own users
and your own environment. Before designing or implementing any service,
talk to your users (and potential users). Consider how they want to interact
with the library and what services they may find useful, and try to deliver
what they need, rather than picking services that you may want to introduce.

One trend is emerging that few users seem to flag as important. 'Second
screening',[2] where users support their main screen (such as a fixed PC) with
a mobile device like a smartphone or tablet computer, is becoming more
common. Users may start to use more mobile content in your libraries,
particularly instructional materials such as podcasts and video tutorials, to
support their main activities. This is a facility that users tend not to ask for,
or to say is important to them, but that is emerging as increasingly common.

There is a danger in taking what your users say in too much detail. Steve
Jobs was famously dismissive of the idea, saying 'You can't just ask customers
what they want and then try to give that to them. By the time you get it built,
they'll want something new.' Not many of us have his vision for innovations,
however, and for most of us focus groups are incredibly useful. Do take the
individual services that they mention with a pinch of salt, however. Don't

slavishly follow any rankings of services that you may get from surveys or focus groups, but instead use the focus groups to collect the much more useful information about how users wish to interact with your services. The most useful information from this study was to do with attitude: students loved being contacted by the Library with information that they saw as being immediately useful (so SMS reminders and similar services were desirable); students had to see the immediate utility of services (so don't spend time on new mobile services that require a lot of user education); and users tended not to think about interacting with most library services unless they were in (or en route to) the Library (so investigate services that are useful within the library building rather than ones that would be used only at a distance).

Acknowledgement

This chapter is an adapted and updated version of an article produced thanks to a Library and Information Research Group award in 2009 and was first published as:

Walsh, Andrew (2010) Mobile Phone Services and UK Higher Education Students, What Do They Want from the Library? *Library and Information Research*, **34** (106), 22–36.

Notes

1   www.alps-cetl.ac.uk.
2   http://trends.masie.com/archives/2011/3/25/657-second-screens-unofficial-learning-devices.html.

References

Gauzente, C. et al. (2008) Attitude toward M-advertising, Perceived Intrusiveness, Perceived Ad-clutter and Behavioral Consequences: a preliminary study. In: *Proceedings of the 19th International Conference on Database and Expert Systems Applications*, 461–5.
Jones, G., Edwards, G. and Reid, A. (2008) Supporting and Enhancing Undergraduate Learning with m-learning Tools: an exploration and analysis of the potential of mobile phones and SMS. In: *Proceedings of the 6th International Conference on Networked Learning*, www.networkedlearningconference.org.uk/past/nlc2008/abstracts/PDFs/Jones_162–170.pdf.
Karjaluoto, H., Leppäniemi, M., Standing, C., Kajalo, S., Merisavo, M.,

Virtanen, V. and Salmenkivi, S. (2006) Individual Differences in the Use of Mobile Services among Finnish Consumers, *International Journal of Mobile Marketing*, **1** (2), 4–10.

Karjaluoto, H., Standing, C., Becker, M. and Leppäniemi, M. (2008) Factors Affecting Finnish Consumers' Intention to Receive SMS Marketing: a conceptual model and an empirical study, *International Journal of Electronic Business*, **6** (4), 298–318.

Merisavo, M., Kajalo, S., Karjaluoto, H., Virtanen, V., Salmenkivi, S., Raulas, M. and Leppäniemi, M. (2007) An Empirical Study of the Drivers of Consumer Acceptance of Mobile Advertising, *Journal of Interactive Advertising*, **7** (2), 1–18.

Pasanen, I. (2002) Around the World to Helsinki University of Technology: new library services for mobile users, *Library Hi Tech News*, **19** (5), 25–28.

Uday Bhaksar, N. and Govindarajulu, P. (2008) Implications of Mobile Phone Technology Usage on Learners in a Learning Process, *International Journal of Computer Science and Network Security*, **8** (5), 251–9.

Further reading

Cassidy, E., Britsch, J., Griffin G., Manolovitz, T., Shen L. and Turney, L. (2011) Higher Education and Emerging Technologies: student usage, preferences, and lessons for library services, *Reference & User Services Quarterly*, **50** (4), 380–91.
A recent study looking at students' attitudes to and usage of various technologies at Sam Houston State University. It covers mobile devices, and also social media and the interactions that students would like to see with the library through social media as well as via any particular type of device.

Folley, S. and Jabbar, A. (2010) *Mobile Learning Project Report*, University of Huddersfield, http://eprints.hud.ac.uk/8815/.
A project report based on experiences at the University of Huddersfield. This report discusses students' attitudes towards the university's using services such as text messaging and podcasting to support their learning. It includes a snapshot of the sorts of devices students used and, most importantly, asked students how they felt about various services that libraries might consider introducing.

Li-Ping Tang, T. and Austin, J. (2009) Students' Perceptions of Teaching Technologies, Application of Technologies, and Academic Performance, *Computers & Education*, **53** (4), 1241–55.
More general, but also more in depth than the other studies here, this article is the result of a series of in-depth studies into how learners use

and experience technology. The ways in which learners select and use appropriate technologies, as reported in this paper, have implications for how libraries should consider introducing mobile technologies.

Parsons, G. (2010) Information Provision for HE Distance Learners Using Mobile Devices, *The Electronic Library*, **28** (2), 231–44.

This study focuses on UK distance-learning students at a Scottish university, as opposed to the more 'traditional' students often surveyed in other reports. It provides a good summary of how these students related to mobile devices and information sources at the time of the study. It also illustrates how difficult it was for these users to think about future use. Almost half of the respondents in the study had bought or would buy a mobile device for education, but they struggled to think of ways in which they would like to interact with information in a mobile environment.

Paterson, L. and Low, B. (2011) Student Attitudes towards Mobile Library Services for Smartphones, *Library Hi Tech*, **29** (3), 412–23.

The results of JISC-funded surveys and focus groups at Edinburgh University in the UK. Interestingly, in two surveys seven months apart, student ownership of smartphones rose from 50% to 67%, with 68% of the remainder intending to upgrade to one within 12 months. The results suggest that the students surveyed would like to see access to library accounts (so that they could check due dates, renew items and reserve items); a mobile-friendly Online Public Access Catalogue (OPAC) that would allow users to search effectively for items in the library while on the move; mobile versions of floor plans and maps of the library; access to live library information, including PC availability; and a booking system that would allow users to make and amend library room bookings from their phones.

Ramdsen, A. (2010) *The Level of Student Engagement with QR Codes: findings from a cross institutional study*, Working paper, Bath: University of Bath, http://opus.bath.ac.uk/19974/.

This working paper shows the level of student awareness and engagement with one particular technology in a UK university that has experimented a great deal with QR codes. It finds similar reservations amongst its student population to those found in the focus groups reported on in this chapter.

W2C (2011) Responses from 100 Device-led Student Interviews. *JISC W2C Project Blog*, http://lrt.mmu.ac.uk/w2c/2011/06/11/responses-from-100-device-led-student-interviews/.

Some results from the W2C project (a JISC funded-project at Manchester Metropolitan University in the UK). The project interviewed 100 students in early 2011 about their use of technology and study habits. It has some interesting results that illustrate the types of devices people own and use when they come into study spaces such as academic libraries.

Young, J. (2012) *Episode 91: students want colleges to go mobile now, even if services aren't perfect*, Tech Therapy Podcast, http://chronicle.com/blogs/techtherapy/2012/01/04/episode-91-students-want-colleges-to-go-mobile-now-even-if-services-aren't-perfect/. Slightly different to most of the journal articles referenced in these lists of further reading, this item is a podcast, part of a series called the Tech Therapy Podcasts that cover tech issues of interest to 'professors, administrators, and students'. This is an interesting interview with Cindy Bixler, who argues that the biggest challenge to mobile implementation in colleges is a tendency to over-plan.

2

Modelling mobile information literacy

Introduction

Information literacy is something that concerns many of us in libraries, whatever our sector. There are many models available to choose from, but I would class most as 'standards' rather than as 'models'. They describe what attributes or competencies the 'information-literate person' should have, implying that it is fairly easy to work out who is or isn't information literate.

Models that take how people experience finding and using information in different contexts, those that model the richness of real behaviour, may be of more use to most of us than these generic, competency-based models that we are used to. Models are available, particularly those emanating from Queensland University of Technology (QUT) and Christine Bruce's research group (Figure 2.1), that have built towards the idea of 'informed learning' (Bruce and Hughes, 2010).

But what about the mobile world? What would a model of information literacy for people searching for, evaluating and using information in mobile

Bruce, C. (1997) *The Seven Faces of Information Literacy*, Adelaide: Auslib Press.
   The first relational model, based on Christine Bruce's PhD research.
Edwards, S. (2006) *Panning for Gold: information literacy and the Net Lenses model*,
   Adelaide: Auslib Press.
   This brings the idea of 'lenses' into information literacy, terminology that has permeated more recent models such as the new SCONUL Seven Pillars model in the UK.
Bruce, C. (2008) *Informed Learning*, Chicago: ACRL.
   This extends the ideas coming from relational information literacy into the wider area of how people use information to learn.

Figure 2.1 *Some major relational information literacy models*

contexts look like? This chapter outlines some starting-points for a model for mobile information literacy. Hopefully, if we can see how people interact with information on the move, we can then start to think more seriously about the sorts of services we should be providing through mobile devices and what training we may need to offer to our staff and users. Together with the previous chapter, describing what mobile users want from the library, this will help to provide context for the examples and ideas in the rest of this book.

Areas of variation between fixed information literacy and mobile information literacy, from the literature

The move towards searching for and using information on the move, from a variety of handheld devices, moves us further away again from the comprehensive, competency-based approach of the dominant information-literacy models. The tools or models that we may need in order to make sense of information literacy within a mobile context will be significantly different to the ones that will fit a fixed information environment. In order to get useful knowledge about how people act in the mobile environment, we really need to consider the unique nature of satisfying information needs on the move.

Based on the existing literature about how people interact with information in a mobile environment, there are four key areas (see Table 2.1) in which mobile information literacy (IL) will vary from traditional views of information literacy, or 'fixed' information literacy.

Table 2.1 *Differences between fixed and mobile information literacy*

	'Fixed' IL	**'Mobile' IL**
Where	Largely in 'set' places. At a desktop computer (with little variation in software); at a fixed workplace; within a library.	Anywhere; any mobile device (phone, games device, e-book reader – massive variation in devices).
What	Anything and everything.	Normally quickly found information, often context or location specific.
How	Range of established tools to access and manage a wide range of information sources. Standard search engines.	Often narrow apps and individual specialist sites rather than open web.
Time spent	Varies. Often slow, long access. People spend long periods searching for, organizing and extracting information, especially for academic use.	Short only. Quicker searches. Little pondering and extracting of information. Favours short chunks of information, 'convenience' of device.

Where it is manifested

Church and Smyth (2008) carried out a diary study of mobile information needs, asking participants to note whenever they had a need to find information. They found that over 67% of information needs might be generated when the user was mobile. The numbers of people who own mobile devices mean that we can increasingly provide services to meet these needs as they occur. Indeed, Heimonen (2009) found that amongst already active mobile internet users, virtually all of these 'on the move' types of information need were addressed through mobile devices as they occurred (145 out of 147 information needs). The only failure to address a need, in this study, was due to a mobile phone battery's having run out.

So mobile search can and does happen anywhere, from a range of devices with massive variations in functionality. Mobile searching can be done from practically any mobile device that has the ability to connect to the internet. People now search for information from mobile phones, PDAs, handheld games devices, e-book readers, tablets (touchscreen portable computers), netbooks, laptops and more; and this can happen in any location with a mobile phone or wireless internet signal. Searching no longer happens in fixed, controlled environments, but in random, messy, uncontrolled ones – from crowded public transport on the way to work, to the loneliness of Mount Everest.

What searches are carried out

Mobile information needs are dominated by the desire for quick access to often context-specific information, particularly regarding local services, travel and trivia (Church and Smyth, 2008 and Heimonen, 2009). Whereas with searching and using information in a fixed, traditional location we may search for anything and everything, this isn't the case for mobile use. The searches we carry out on a mobile device are much more likely to be an additional activity, rather than the sole focus of our attention. They are therefore influenced by the primary activity that we are also engaged in: in other words, the context in which we find ourselves (Hinze et al., 2010).

The information we seek while on the move is facts and small elements of information. We look for the time of the next train, the way to the station, perhaps the closest place to eat while we are waiting, not for discussions about the reliability of trains, the reasons why a train station is located in a particular place, or the place of take-away cuisine in our cultural heritage. There is likely to be limited evaluation of the information we find, and little opportunity to take away detailed information and to derive new knowledge from it. Detailed information is to be avoided as being hard to read on the small screens we may be using, or too time consuming to look at in this context.

How we search

Using a fixed, fully powered computer allows access to a wide range of established tools and information sources. It is typical to start searching for information with a generic search engine, and this may then lead on to more specialist sites or search tools. Searching on a fixed computer could be characterized by the breadth of sources and tools available and used.

Mobile searching, however, is heavily influenced by the natural constraints of using a device with a small screen and a small or virtual keyboard, and may be characterized by the narrowness of the sources used. Mobile phone users are often encouraged to access the web through their mobile network's own portal – which, until recently, were the most-visited mobile websites (Church and Smyth, 2008) – and through that to access only a small part of the mobile web.

Time spent on searching

Kamvar et al. (2009, p. 805) found that less time was spent refining searches on mobile phones, with 1.94 queries per search session on a desktop and 1.70 from a typical mobile. There is less patience when searching from a mobile device and the desire for quick and easy searches dominates.

In Heimonen's (2009) study it was found that 35% of information needs occurred in the home. Even though a fixed computer (or laptop) may have been available, the speed, proximity and convenience of using a mobile device trumped the more powerful device.

It can be said that people turn to their mobile devices for 'quick and dirty' searches for information. They want to know something, and they want to know it as quickly as possible.

So what does this mean?

There is a desire for information on the move that is often context or location specific. With the recent increase in the number of internet-enabled mobile devices, from iPhones to e-book readers, it is becoming the norm to possess a device that we can use to try to satisfy those information needs and desires.

The current dominance of competency-based information literacy models does not take into account the changing nature of information discovery and use on the move, and never could, due to the rapidly evolving technological landscape. Instead, the dominant models are largely lists of 'standards', and ignore true information literacy models that describe how people act in real life. In the absence of such models, we therefore need to consider the areas discussed here in order to adapt our practice to support the reality of everyday interaction with information on the move, rather than describe an 'ideal' information-literate person whom we may wish for and may never meet.

Also, in an increasingly mobile world we shouldn't assume that our users will continue to interact with information in the same way that we feel they always have. We already lag behind reality in certain ways, designing our physical libraries in ways that made sense in a largely paper-based world and failing to make electronic resources easily discoverable in the same environment. It is important to consider what it means to be information literate in a largely mobile world, so that we can see what services and tools make sense to our users.

Fuller models of mobile information literacy are in development and, from interviews with confident mobile users, I have confirmed that people want searching to be 'quick and easy'. Their mobile information needs tend to be highly context specific, influenced by the environment that they are currently in. They often use social networks to discover information, making searching a social activity. They use mobile devices and the ready access to 3G and Wi-Fi networks to move information rapidly between devices, often searching on one device and consuming content on another. Alongside this they treat their mobile devices as an extension to their own memory, storing information 'in the cloud' by using mobile devices or relying on quick searching to discover information that otherwise they would have memorized. Alongside all of this, many people struggle to cope with the information overload engendered by always being connected to the online world through mobile devices.

A single quote that nicely sums up the degree to which constant mobile access to information has changed behaviour came from an interviewee in my own research:

> '... it has become like an outboard brain for me, you know, and it in a way is kind of an extension of me. It is interesting that having something like this will allow you to kind of delegate remembering facts and free you up for kind of critical thinking. ... I organize everything through it, you know, I access all my e-mail through it, I use it to keep up with my work here, I synch up with my work calendar, my personal Google calendar, to schedule everything, I keep recipe notes on it and things like that, it's just all ... it's everything. Everything that I, that is my personal information that I use on a daily basis, is either on this or accessible from it.'

Summary

This chapter sums up some of the differences between the varied ways in which people interact with information in a mobile world, but there is much more to come out of current research. In the meantime, we should take these descriptions of how people interact with information in a mobile environment and use them to inform how we develop mobile-friendly library services. If people want searching for information to be 'quick and easy' on a mobile device, then there is little point in replicating traditional library databases in a mobile environment. If people struggle with the constant push of information to their mobile devices, then perhaps we should be thinking about what help and training we might be able to offer in order to deal with that influx of information. When, often, we find, from talking to users, that mobile searching can be a social activity, we should consider building recommendations into mobile search results based on the usage history of all our users.

Combine the information from this chapter with the material on what our users want from mobile library services (Chapter 1) in order to decide which of the many possibilities and ideas for mobile services are right for your users, and to consider what will make the greatest future impact on your library's users.

References

Bruce, C. and Hughes, H. (2010) Informed Learning: a pedagogical construct attending simultaneously to information use and learning, *Library and Information Science Research*, **32** (4), A2–A8.

Church, K. and Smyth, B. (2008) Understanding Mobile Information Needs, *Proceedings of the 10th International Conference on Human–Computer Interaction with Mobile Devices and Services (Mobile HCI '08)*, New York: ACM, pp. 493–4.

Heimonen, T. (2009) Information Needs and Practices of Active Mobile Internet Users, *Mobility '09 Proceedings of the 6th International Conference on Mobile Technology, Application and Systems*, 10–13 September, Nice, France.

Hinze, A. M., Chang, C. and Nichols, D. M. (2010) *Contextual Queries and Situated Information Needs for Mobile Users* (Working paper 01/2010), Department of Computer Science, University of Waikato, Hamilton, New Zealand, www.cs.waikato.ac.nz/pubs/wp/2010/uow-cs-wp-2010–01.pdf.

Kamvar, M., Kellar, M., Patel, R. and Xu, Y. (2009) Computers and iPhones and Mobile Phones, Oh My: A logs-based comparison of search users on different devices, *WWW '09 Proceedings of the 18th International Conference on World Wide Web*, New York, www2009.org/proceedings/pdf/p801.pdf.

Further reading

Wright, S. and Parchoma, G. (2011) Technologies for Learning? An actor-network theory critique of 'affordances' in research on mobile learning, *Research in Learning Technology*, **19** (3), 247–58.

Getting into more depth on the ways in which people use mobiles to learn, this article presents a serious theoretical framework for mobile learning that can serve as a useful jumping-off point for the kinds of theories that it may be useful to consider in mobile information literacy.

Yarmey, K. (2011) Student Information Literacy in the Mobile Environment, *EDUCAUSE Quarterly (EQ)*, **34** (1), www.educause.edu/EDUCAUSE+Quarterly/EDUCAUSEQuarterlyMaga zineVolum/StudentInformationLiteracyinth/225860.

Describes how University of Scranton students search for, evaluate and use information using mobile devices. The study was an electronic survey, so doesn't pull out some of the rich variation of use that interviews may do, but it gives an interesting idea of how student behaviour is changing. It also gives some recommendations as to how libraries should change their behaviour so as to help students to develop appropriate information literacy in this environment.

3

The mobile librarian

Introduction

This chapter looks more at how we, as information professionals, can become more mobile than we may have been before. Rather than considering how mobile devices can be utilized by our library users to access our services, it looks at how we might use them to become more mobile ourselves.

First of all, this chapter covers how we can use mobile devices to support our users while on the move, particularly through the idea of turning ourselves into roving librarians. The available technology has reached the stage where many of us can effectively take a large proportion of our library's resources with us in a device that can fit in one hand, so we have great opportunities to take library support to our users, wherever they may be.

The chapter then moves on to discuss how we can make the most of mobile devices to be as productive as possible, using a range of tools and devices in order to become more productive wherever we are. It can be a struggle to find time to sit behind a desk and computer to work. Mobile technology can help us to work wherever we find ourselves, on whatever piece of technology we have to hand – as long as we know how to take advantage of it effectively!

Finally, the chapter covers how mobile devices can help directly to support our continual professional development, helping us to keep up to date and to carry out research. There are many mobile-friendly tools that can help to turn otherwise wasted time into an opportunity to develop ourselves. This section covers a few examples of these tools.

Mobile support and reference

A core part of many librarians' role is to offer reference-desk types of support to library users. This might be at a traditional reference desk, in a back office or room used for longer appointments or at the desks of other staff and researchers. Normally, however, it will be at a desk. It will be at fixed locations that you and the person you are helping will need to go to. If someone asks for help while we are elsewhere in the library or the institution where we work, then most of us will tend to encourage the questioner to come to that fixed space if we need to show them any online materials or to look up any facts that we don't immediately know.

Mobile devices, particularly tablets and phones, are ideal for offering this sort of support where your users require it and not solely at these limited, fixed locations. They can provide you with alternative methods of support, whether these replace your normal desk or supplement it. They allow you to be considerably more agile in answering your users' needs and to improve your service to them.

Many libraries, particularly public and academic libraries, have implemented roving support for their users. This is often IT focused, but can also cover a range of other support. Many case studies have been published showing that well-planned and implemented roving support has benefits.

If you already have staff offering roving support, the addition of a device such as an internet-enabled slate or tablet computer (for example an Apple iPad) can extend significantly the support they can offer to people on the spot. Besides easily replacing any paper lists of frequently asked questions, a tablet computer will connect to the same information sources to which a fixed desk computer may have access. It will allow you to replicate much of the support that you would offer through a fixed desk, while on the move. The larger screen of a tablet computer, as compared to a phone or PDA, also enables you to show the questioner how to resolve their problem or search for information. It helps to move your roving staff from *describing* how to resolve a problem to *demonstrating* it on the spot, wherever they are asked the question. For example, libraries such as the University of Northern British Columbia's Geoffrey R. Weller Library use iPads to provide roving support during peak periods.[1]

Mobile devices can also turn many more of your staff into roving support outside as well as inside the library. Equipped with a smartphone, or preferably a tablet, a librarian can give support when and where it is asked for. A librarian at the University of Queensland recently described to me how she turned a meeting with an academic into an impromptu support session. Equipped with an iPad, when asked a question she was immediately able to demonstrate library resources to the academic without needing to make a meeting appointment in order to have access to the resources via a fixed computer. The level of service was significantly better than she could have offered without the mobile device. In academic, health and specialist libraries, where librarians may often be performing outreach work, mobile devices enable us to carry our libraries with us, turning every librarian with a device into a freely roving reference desk.

A mobile device can also enable us to 'set up shop' where our potential users are, rather than with the users we've already managed to get through the doors of our library. A decent mobile device makes outreach work much easier. At my institution we are increasingly going into the students' union and into informal spaces within our academic schools. Initially we tried this with laptops because the whole campus is covered by a Wi-Fi network, but they are heavy, have limited battery life and are awkward to demonstrate resources on. Holding a heavy laptop vaguely pointed towards a student and trying to search online library resources while standing in a busy canteen is not to be recommended! Using a relatively light tablet computer with a long battery life is much easier, both for the librarian and for the people you are helping. A tablet computer makes it easy to go into spaces where your potential users are and for you to offer direct support that these users probably wouldn't otherwise receive.

At the University of Huddersfield we've recently equipped all our subject specialist librarians with tablet computers (a mixture of iPads and ASUS Transformers) and are regularly visiting student spaces to offer support. This has received great initial feedback. My favourite quote so far from our experiments with roving librarians (see Figure 3.1) comes from an applied science student whose query was answered at exactly the point when they needed us: 'It's like you were sent from God!' While most of the feedback we receive isn't so over the top, it is overwhelmingly positive.

Sending roving support directly into informal spaces, impromptu teaching in unexpected places and somewhat more formal drop-in sessions can all be enabled by embracing these mobile technologies. They are a real boon for outreach work and allow you to reach people who might otherwise never come and ask you for help.

Finally, for mobile support and reference, consider using mobile devices as a way of being virtually present when you cannot physically be so. Services such as Facetime (on Apple iOS devices) and Skype (on most platforms) can be combined with your Wi-Fi service and mobile

Figure 3.1
Roving Librarian logo from the University of Huddersfield

device to create free, unlimited video calling. Some Scandinavian libraries have already started experimenting with this idea in order to staff reference desks in their branch libraries. Instead of having a librarian in each branch library, they instead secure a tablet computer in a suitable location within the branch. This has a permanently open connection to a librarian sited in a larger library, allowing one librarian to staff many library reference desks. With the large, clear screens of the tablet computers and free, reliable video conferencing software, the users can feel as if they are looking through a window to the librarian, rather than being many miles away.

Mobile productivity for librarians

Whatever type of library they work in, librarians are increasingly expected to move beyond the confines of their physical libraries. Public librarians may take part in outreach work in their communities. Health librarians may meet clinicians and go onto the wards with consultants. As an academic librarian, I seem to spend a great deal of my time in meetings with my academics, away from the library. Then there is also the time we may spend travelling, or at home after work officially finishes, either working or catching up on our own professional development.

This loosening of the ties between our desks (and fixed computers) in the library and where we actually carry out work can cause difficulties. We can't access the pile of paper from our desk when we are meeting a consultant on the other side of the hospital. We can't open up a file stored on our work computer's hard drive from home. We may have information stored in many different places and on several different devices (work, home, laptop, fixed computer(s), mobile phone) and struggle to access the right piece of information at the right time.

There are tools and techniques that can help us. Taking advantage of some of them can help us to share information across fixed and mobile devices and use our time more productively.

The first type of tool uses the internet as a giant drive for documents and files. The example of Dropbox is given below, but Apple (iCloud), Amazon (Cloud Drive, which it also sells through many third-party operators) and Microsoft (Skydrive) are all major options. These tools allow you to save a file on one device and continue to work on it from any other device capable of connecting to the internet. You can save a document on your work computer, read through it on your mobile phone or tablet on the way to a meeting, display it on another device in the meeting, perhaps editing it as you receive feedback, and know that the same document will be available for you to finish off later, working on your work or home computer. For me, these drives have

Dropbox

Dropbox (www.dropbox.com) is a good example of a 'cloud' drive. You can access it from Windows-, Mac- and Linux-based computers as well as from Apple's i-devices, from Android devices and from Blackberry phones.

On a computer it shows up as a folder that you can drag and drop files into or save directly to, just as you would with any folder. It then automatically synchronizes those files with a version on the internet.

At the time of writing, Dropbox works on a 'freemium' model, offering 2GB of online storage space for free, with increased space available for a monthly fee.

completely replaced USB drives, giving me increased flexibility in how and where I access my files and making me much more productive than before. This book has been written with the help of one of these drives, allowing me to work primarily at home, but also to edit and write chapters using my netbook on the train and in hotels, or to read through case studies on my mobile phone wherever I may be.

There are security and privacy implications to using these drives. If you want to store confidential or sensitive information, consider services that encrypt your data and do not claim ownership over anything stored on their cloud drives. If security and/or privacy are an issue, make sure that you check the terms and conditions of any cloud services before using them.

Many of these drives allow you to share folders with others, so that documents can be shared and worked on collaboratively with colleagues, which leads us to the next type of tool, online word processing and other office-type software. These tools also normally include space on which to store the documents you create and to share them with others.

The best-known software for word processing, creating spreadsheets, presentations and more on fixed devices is Microsoft Office. There is an online version of this, Office Web Apps, which is available through many routes. It works directly within Microsoft offerings such as its Skydrive or through routes such as Docs.com, which integrates with Facebook messages.

In the online world, however, Google's offering, Google Docs, has become the best-known online collaborative system for working on documents. Google Docs works with a range of formats, so it is possible to upload documents originally created using other office software. Once a file is created or uploaded, it can be shared easily with everyone on the web or with nominated individuals, so that it can be worked on collaboratively over the internet on anything from a fixed computer to a mobile phone.

There are many other applications with similar functionality, such as Zoho (www.zoho.com, currently with free and paid-for versions), but the two major ones are Office Web Apps and Google Docs.

Sharing is not limited to files and documents. Sharing of ideas is also something that mobile devices, able to connect to the internet from practically anywhere, can really help us with. Many of us build personal and professional learning networks of people – particularly using social media – that we can interact with and that help us to build our professional

Google Docs

This offers a word processor, spreadsheets, presentations and more. It is becoming increasingly popular to use a Google Docs spreadsheet with the 'form' functionality to create free online surveys.

It is freely available to anyone through https://docs.google.com, requiring only a Google account. The mobile version works well with a range of devices, allowing you to create, view and edit documents on the move without installing software.

knowledge. We can follow conferences from a distance via Twitter (www.twitter.com), share presentations on SlideShare (www.slideshare.net) or get involved in professional discussions on Google Plus (https://plus.google.com). Personally, I find it difficult to spend time on these networks while at my work desk. I also know many people, particularly in health, school and public libraries, who have access to these networks blocked from their work computers. Mobile phones can be the answer to these problems of time and access to useful tools that allow the building of valuable learning networks. You can sidestep the problems of access over a restricted work network by using your own mobile network whenever necessary.

Many social media tools work well over mobile devices, and this is the main way that I use tools such as Facebook (www.facebook.com), Twitter and Google Plus to build and participate in my own personal and professional learning networks. I access these tools primarily while on the move, which allows me to snatch back time that would otherwise be wasted in transit and to use it instead to build my professional knowledge. Without access to these

Google Plus

(https://plus.google.com) This is Google's latest attempt at a social network and it extends into many of its other services, including search results – which has caused some fuss over the increasing influence of social search.

At its most basic level, you can follow posts (text, photos, videos, links, check-ins) created by other people in your network, leave comments on those posts, +1 posts to show that you agree with them, and share other people's posts, in addition to posting your own updates.

A major difference to other social networks is that you place people into 'circles'. When sharing information, you can then choose which of your circles can see that information.

tools through my mobile phone, 1 would have considerably poorer professional networks and would find it much more difficult to keep up to date with developments in my areas of interest.

Keeping up to date and research tools

Keeping up to date by using social media to build personal and professional learning networks has been looked at above, but what about more formal ways of keeping up to date?

Many of the tools that you may be familiar with for research and keeping up to date via fixed computers can now be used on mobile devices, such as tablets and mobile phones.

RSS (Real Simple Syndication) feeds are often generated by online tools that we may be using for our professional development. Many websites, news sources, blogs and subscription databases will provide RSS feeds for new content. It is often possible to create RSS feeds of searches that we have performed for content from many different sources, and these automatically update us with new results for those searches. Typically, people collate all of their RSS feeds into a single reader that they access from a desktop computer, and they often struggle to keep up to date with the quantity of feeds that they have set up. However, RSS readers also work well on mobile devices, and keeping up to date in small chunks of time, using a mobile device, is an ideal way to cope with the quantity of information that these feeds can produce. Instead of relying on having time to check an RSS reader at home or in your office, you can spend a few minutes at a time on a mobile device while also engaged in other activities.

Traditional library databases are increasingly being offered in mobile formats, letting us keep up to date on research wherever we are. Aaron Tay (Librarian at the National University of Singapore Libraries) carried out a survey of such databases, and found a mixed set of results.[2] They tended to have limited search options and sorting of results, as well as limited options for transferring results between devices – something that is important if these tools are to become really useful to us in the future. There are also problems with authentication in mobile apps (though not normally in mobile websites), meaning that you may need to set up different codes and passwords in order to gain access to institutional databases.

This reflects the emerging nature of mobile-friendly databases, and they are likely to be better and easier to access and make use of as they become more established. At the moment, mobile websites are likely to provide a better experience than apps, especially with regard to institutional logins. Potentially, however, apps could more easily facilitate the sharing of results across different devices, which is important for saving us time when on the move.

An area where mobile-friendly services have already started to make an impact is in 'next generation' catalogues and 'discovery services', where new products are being created rather than existing products having mobile interfaces built on top of them. These products often include mobile web interfaces as a standard feature, providing a good user experience and allowing us to check our library's holding from a range of mobile devices.

Being able to keep up to date by scanning through RSS feeds, or to use mobile tools to do a quick search for a book or journal article while on the move, is a great thing to be able to do. But it is only the start of any process of professional development. While finding material using a mobile device is a useful time saver and convenient for many of us, using that material is another matter. Once you find something of interest, you then need to read it, perhaps make notes and probably save it somewhere for a later date. There are mobile tools that can help, though at the moment it is sometimes difficult to tie them together.

The most important set of tools may be reference management software. If you can find useful references from your mobile device, a piece of software that can manage them for you and make them available across all your devices, including desktop computer, can help you to return to them and read through them in detail. The large commercial providers of such software, such as Endnote and Refworks, normally make mobile versions available (both of them through mobile web pages), though free alternative systems such as Mendeley are also starting to appear in mobile versions (a Mendeley app is available for Apple devices) that synchronize with your desktop equivalent. There is also a plethora of device-specific apps that use standards such as BibTex to allow you to manage references, though you would need to copy the files to and fro between devices to make them usable across platforms.

Reference managers now tend to include the option to create notes, including directly onto PDFs of full-text articles, and this can be incredibly useful when extracting information to keep up to date. This functionality isn't normally available in mobile versions of the software, so you may need to look for alternatives. There are many note-making systems available, but one of the best known that works across all platforms is Evernote (www.evernote.com). It allows you to make notes, save images and snapshots of websites and organize them in a format that you can access from desktop or mobile devices. While it is difficult to organize long lists of references in systems such as Evernote, it does provide a much richer and easier environment in which to make notes while on the move and revisit them from a desktop computer.

Summary

Mobile devices can allow us to be more mobile than ever before in our delivery of library services. We can often carry significant proportions of a library's stock with us on mobile devices, through the internet connectivity they provide. They make it easier for us to help our users outside the normal constraints of the library.

Beyond this, mobile devices also make it easier for us to keep up to date and continually develop professionally. They allow us to squeeze productivity into time that otherwise wouldn't be used, by letting us work wherever we are.

Both of these elements require a certain change of mind-set in order to work. We need to see time spent waiting for meetings to start, time spent on public transport, or even those few minutes walking from place to place, as opportunities to use our mobile devices for our own development. Equally, we need to see wherever our users happen to be as places where we can offer support. With mobile technologies, we no longer have to direct users to the library: we can bring the library to our users.

Notes

1 http://journal.code4lib.org/articles/5038.
2 http://musingsaboutlibrarianship.blogspot.com/2011/08/what-are-mobile-friendly-library.html.

Further reading

Cheetham, W. (2007) From Table to Tablet – How a Wireless Tablet PC Could Help to Deliver a New Customer Service Paradigm in a Public Library, Conference paper from *ALIA Information Online conference, Sydney, January,* http://conferences.alia.org.au/online2007/Presentations/1Feb.A18.from. table.to.tablet.pdf .

A case study of how a tablet PC allowed an innovative Australian public library service to introduce a mobile reference service, and how this was received by staff and users.

Dempsey, M. (2011) Blending the Trends: a holistic approach to reference services, *Public Services Quarterly*, **7** (1–2), 3–17.

In addition to the case study of a library's introduction of a roving librarian service, this article helps to put the roving service into context with other reference services, offering a blended model of support to library users.

Hall, E. and Gomes, A. (2011) Going Mobile: a medical library's outreach strategy, *Journal of Electronic Resources in Medical Libraries*, **8** (3), 234–42.

This article describes a range of activities undertaken to make librarians more mobile in a medical library setting, in addition to making more mobile-friendly services available to users.

Lippincott, J. (2010) Mobile Reference: what are the questions? *The Reference Librarian*, **51** (1), 1–11.

An overview of some of the factors libraries should take into account when considering offering mobile reference. Rather than going into depth on any particular project, this paper gives a considered overview that could help a library service to plan the introduction of mobile reference services.

McCabe, K. and Macdonald, J. (2011) Roaming Reference: reinvigorating reference through point of need service, *Partnership: the Canadian Journal*

of Library and Information Practice and Research, **6** (2),
http://journal.lib.uoguelph.ca/index.php/perj/article/viewArticle/1496.
A Canadian case study showing how the University of Northern British
Columbia addressed declining use of traditional reference services by
introducing a roving reference service. Staff used iPads to leave the
safety of the reference desk and go to students where they studied.

Penner, K. (2011) Mobile Technologies and Roving Reference, *Public Services
Quarterly,* **7** (1–2), 27–33.
A call for the use of more mobile technologies in order for library staff to
support their users, including a description of how the author used such
technologies in her own library setting.

Reynolds-Pope, T., Chesnes, M. and Early, C. (2010) The Mobile Librarian
Program at the NASA Goddard Space Flight Center, *Science & Technology
Libraries,* **29** (4), 267–75.
This well-described and justified mobile librarian project used laptops to
make the librarians more mobile around the NASA Goddard Space
Flight Center. It is easy to see how the experiences described in this
paper could transfer across to an even more mobile service using tablet
computers.

Widdows, K. (2011) Mobile Technology for Mobile Staff: roving enquiry
support, *Multimedia and Information Technology,* **37** (2), 12–15.
A practical description of how the University of Warwick library in the
UK used smartphones and iPads to provide a roving reference service
within the library.

4

Texting in libraries

Introduction

Of all the mobile-friendly technologies and services that we may consider, text messaging, or SMS, must surely be the first. Even if you believe that your users may not want or be ready for any mobile service that requires a smartphone, you should still be considering text messaging. In a world where the vast majority of people own a mobile phone of some description, our users will be sending and receiving text messages. SMS is the most widely used item of functionality for most mobile phone users, so it seems sensible to take advantage of it.

There is one key danger in using text messaging to deliver library services. The very personal nature of mobile phones can mean that text messages can easily be viewed as intrusive. Any service that you deliver through SMS must therefore meet a real need. Do not use it to promote your services, or to send out regular, indiscriminately targeted messages. Your users will soon have had enough and will object strongly to this abuse of the facility. Instead, deliver services that your patrons will recognize immediately as being to their advantage. It is easy to get carried away, particularly if you have a contract that allows unlimited text messages for a fixed price. My children's school initially got too enthusiastic with its text service, sometimes sending several text messages a day, normally on minor matters, to all parents. This is the fastest way to lose goodwill and to encourage people to opt out of receiving messages.

Alongside this, make it easy for people to opt out of receiving messages from you. Many organizations ask people to opt in to receiving text messages, which

> **SMS (Short Message Service)**
> SMS, widely referred to as text messaging, has been available since the early 1990s and there are few mobile phones that cannot send and receive text messages. It allows messages of up to 160 characters to be sent between mobile phones.

may work for you if you are using the technology for teaching or for one-off services. If you wish to use SMS regularly, perhaps for overdue notices, then it is easier and more effective to tell everyone who uses your service that they will automatically receive text messages unless they opt out. Either way, how to opt out, in other words, how to stop receiving text messages from the service, should be easy and transparent.

Sending messages without a phone

Most of us think of using a mobile phone to send or receive a text message. It is not the only option, however, and certainly not the main option if you wish to use text messaging for your library.

First, for low-volume or one-off uses, you may be able to use free services to convert e-mail messages to SMS. These often require you to know what mobile operator a particular number uses. Many mobile operators will allow you to send an e-mail to a phone number (e.g. 0123456789) followed by the correct suffix provided by the operator (e.g. txt.operator.net). So, in this fictional example, a short e-mail to '0123456789@txt.operator.net' would be sent as a text message to that number. You can often find the correct suffixes to add to phone numbers on operators' websites, though a good place to start might be Wikipedia, which has partial lists for many countries.[1] This has the obvious drawback of needing to know which operator each of your patrons uses in addition to their phone number, but might be a good option for 'one-off' uses.

The main way that most organizations send and receive text messages is via a subscription to a web-based SMS portal. These services allow the sending of large numbers of messages at a time and normally include tools that enable you to create and maintain groups of users, schedule messages in advance and receive messages from your users. They will normally allow mass importing of phone numbers from other records that you may keep.

Typically, these services will be cheaper than a traditional mobile phone contract when you are sending large numbers of text messages, and you can expect the cost of sending or receiving each message to drop if you subscribe to a package that allows larger numbers of messages. The service may come with a standard-looking phone number on which to receive text messages, but also often provides the option of using a shorter phone number that is used together with a keyword. The keyword would need to be the first word of any SMS sent, so as to make sure that the message was received in the right 'pot' within the text portal. So, for example, a 'text a librarian' service might use the Keyword LIBRARY. To ensure that this is received correctly, a typical question might read: 'LIBRARY What time are you open until tonight?'. This may be familiar to people who have voted or taken part in quizzes while watching television programmes, where you send a text message containing option A, B or C to answer a question. Libraries, however, wouldn't normally use a 'premium' number, but one that charged users their normal network cost for sending a text message.

These text portals often provide the option of responding automatically to text messages, so you could set up appropriate messages that people would receive in response to their texts, if you so wished. For example, if you had a dedicated code for enquiries about opening times the portal could be set in advance to respond automatically with that day's opening hours.

Messages from your library management system

Libraries have gradually moved away from sending out messages in paper form. It is slow and expensive to print off reminder notices about overdue items, requested items that are ready to be picked up and similar standard messages. It is fairly standard now to send such messages by e-mail wherever possible, saving staff time and postage charges, as well as getting the notices delivered to users significantly faster.

A small step on from e-mail reminder and other standard messages is to use text messaging. These messages are viewed as being of direct usefulness to our patrons and few people object to receiving them by mobile. Even in countries where there is a cost implication to receiving a text message, the cost can easily be outweighed by the reduced likelihood of receiving fines for overdue items. People are more likely to notice a text message and will respond more quickly than to an e-mail. In my library service, many students rarely check their 'official' e-mail addresses to which we send notices (despite a great deal of encouragement to do so), but will check their mobile phones obsessively.

After a slow start, increasing numbers of library management systems (LMS) have included generation of text messages as one of their standard options. If this is not an option in your LMS, but e-mail messages are, you may be able to get an in-house developer to write a short piece of code to automatically match e-mail addresses to any mobile phone numbers that you hold. The messages can then be imported into and sent from a standard web-based SMS portal.

The advantage of proper integration of text messaging into an LMS, rather than of trying to add it as an extra step, is that options such as renewals by text message are much easier to introduce. The logical next step after sending overdue messages (or underdue messages – sent the day before an item is due back) is to allow users to reply to the text message to request renewal of those items. This is hard to implement unless integrated properly into your LMS, though it is possible with some development work.

Case study: Communicating with students (using 160 characters or fewer)

Anne Mary Inglehearn
Librarian, Leeds College of Building, UK

In 2009, at an MMIT (Multimedia Information and Technology group) conference in Liverpool, I attended a seminar by Russell Prue. He told us, in no uncertain terms, that we should use a method of communicating with students that they chose, rather than imposing our own current routines on them. Radical! He was talking, of course, about text messaging. By coincidence, when I returned to college the IT manager approached me and asked if the library service would take part in a pilot study using a text messaging service provided by Txttools. To cut a long story short, I took him up on his offer. The pilot received excellent results and Leeds College of Building now has an annual licence and the texting service is used across all departments.

Txttools is an online service that we can access from anywhere using a College account that allows us individual log-ins for different departments. Although the service incorporates many different functions, we in the library service use the package in a very simple way. Using a report from the LMS that lists students who are due a first overdue letter, we select text recipients from a 'phonebook' that has been uploaded centrally. Next, we choose a message template and signature from a selection that we have created. The resulting message can be up to 160 characters long, like a regular text, and arrives at a student's phone almost immediately, for far less than the price of a second-class stamp. Until recently, we disabled the reply function and instead asked students to call us if they wanted to renew their loans. This decision was based on the fact that we wanted to retain a human element in our contact and, being a small college, were well able to cope with the additional phone calls. Also, texts were used only for the first overdue notification, because for the second notification we wanted to increase the formality of the warning by using a letter.

Feedback from students has been universally excellent and when we send out a batch of texts we sit back and wait for the phone to ring. In the early days students would often respond with comments like 'thanks for letting me know', 'did you send me this text?' or 'that's really clever'. Today students are more used to the proliferation in the use of text messaging for information and are less surprised to receive a text from the Library. However, we still occasionally find that after we have sent out texts a student actually working in the Library will walk up to the counter and say 'You've just sent me a text, can I renew my books while I'm here please?' This is one of the reasons why we chose to allow students to use their mobiles in the Library, because it seemed hypocritical to ban them when we were actively using this method of communication. Anecdotal evidence suggests that students respond more positively to a text, perhaps because of the perceived formality of a letter.

We have recently considered ways in which we can increase our use of Txttools and will begin by piloting an updated overdues template that allows students to respond to a text and request a renewal. Library staff have some concerns about this, as it will be an extra task in their already busy daily schedule. They also feel that when students call the Library to renew their books it's a good opportunity to explain the library service's rules and procedures regarding loans. However, as we haven't tried this before there's only one way to find out if it will work for us, and that's by having a go! We have also incorporated texting into our student inductions. At the end of each session we invite students to take part in a competition using their phones. To enter, they must text a keyword (which separates responses in the in-box), followed by their answer to a question and their student ID number, to a phone number displayed on the classroom screen. When the competition closes at the end of the autumn induction period all entries will be downloaded from Txttools to a spreadsheet and sorted to show those that were correct. From these a winner will be picked at random and presented with a £25 Amazon voucher. 'Who wants to be a millionaire?' it is not, but we hope that it will encourage some interaction with the Library and prepare students to use this method of communication with us in the future.

How much does it cost? Well, luckily the licence is currently paid for from a central budget, so it costs the Library nothing! This is because text messaging's wide use across other departments would make a shared cost hard to calculate. However, looking at the cost per text across the whole College, it works out at mere pence. Our licence also includes access to free training and an excellent helpline.

All in all, I would recommend using texts to other libraries, as it's a quick, cheap and efficient way of keeping in touch with your users.

Text a librarian and more

Most of us allow many different ways of contacting our libraries for help and support. Few, however, include sending a text message as a normal option.

If your library service currently has no other mobile-friendly services, I would recommend that you start with a 'text a librarian' type of service. It can be quick, easy and cheap to introduce such a service and it is an ideal entry into the world of providing services via mobile devices.

It is possible to introduce a text-a-librarian service using a standard mobile phone. Indeed, I know of one library that took its director's work mobile and put it in the desk drawer of the enquiry desk. Nothing more was needed, except to promote the number! Preferably, however, you should use an SMS portal, as described above. Not only will this give you more flexibility than a 'real' phone (messages can be answered from any computer), but it will allow you to experiment with web-based SMS portals and may then extend into other services.

Once you start with one text service, you may find many other uses too. In my Library's text service we have one short text number with numerous keywords to direct the messages to different parts of the service. We have keywords for subject enquiries, IT help and disability support. Our most popular service has the keyword 'NOISE'. If users send a message to our standard number containing 'NOISE', followed by a description of where they are in the library, our staff will swiftly investigate the area for anyone misbehaving. This effectively allows the anonymous reporting of bad behaviour and has been repeatedly praised by our students as a great service.

Teaching using SMS

Mobile phones can be used in many different ways when teaching information skills. This is covered in more detail in Chapter 7, so just a few brief suggestions are included here.

As with many other services into which you can introduce mobile phones, text messaging is an ideal first step when using mobiles in teaching. Using text messaging and people's own mobile phones, there is no significant learning curve, the cost implications are extremely low and the vast majority of people whom you may need to teach can participate.

You can use text messaging to turn mobile phones into user-owned voting pads or 'clickers'; to send out a series of text messages to supplement inductions or to remind users of key details after their formal teaching session; or perhaps to set up SMS quizzes to check knowledge, using the keywords and automatic responses built into most SMS web portals.

Case study: Using Poll Everywhere in face-to-face sessions
Dina Koutsomichali
University of Brighton, UK

Introduction

One of the biggest problems when teaching information literacy in higher education is engaging individual students and getting them to participate in an active way during face-to-face sessions. One way of addressing this problem is to invite student input through audience participation or class performance systems using handheld devices like clickers (Deleo et al., 2009; Fies and Marshall, 2006; Lantz, 2010). However, these involve lengthy preparation to set up and use the hardware and software (Kay and LeSage, 2009). More recently, web-based systems like Poll Everywhere have appeared, allowing students to participate by text or online (via Twitter or a web form) but without the need for extra hardware or software (Sellar, 2011). This case study looks at using Poll Everywhere in face-to-face sessions conducted with first-year Arts and Design students at the University of Brighton over a period of six months between October 2010 and March 2011.

Poll Everywhere

Poll Everywhere allows the creation of free-text and multiple-choice polls. Slides can be embedded into a PowerPoint presentation and responses are displayed live. Although it requires registration, it is free for audiences of up to 30 participants. There are also various fee-paying plans for larger audiences and educational institutions (see www.polleverywhere.com).

The case study

Poll Everywhere was used in a) inductions to the Library and its resources at the beginning of the year and b) sessions using library resources for student projects or assignments to same-module groups. Groups ranged from to 10 to 120, depending on whether they were inductions in a lecture theatre or smaller groups of students on the same module, working on a particular project or assignment. Free-text, 3-item multiple-choice and yes/no questions, as well as one-sentence gap-fills, were used. An effort was made to create simple and clearly worded questions and response options (Klaas, 2003). Time limits for each poll were two to five minutes, depending on the nature of the activity. The polls were spaced out at intervals over the duration of each session in order to maintain interest. All the polls were embedded in a PowerPoint presentation that was later uploaded to the students' course module in the virtual learning environment (VLE) (Blackboard) along with the students' contributions. Feedback on using the polls and the activities was gathered orally during sessions and through a free-text poll at the end of each session.

Table 4.1 shows some of the activities for which polls were used.

Findings

Creating an account and setting up the polls was problem free. It was also easy to make changes to polls, although sometimes 'keywords' or 'codes' for multiple-choice questions were not available because they were already in use by someone else.

Polls can be reused by deleting previous answers and restarting them. This saves time and means that old polls can be reused very quickly during a session.

Polls proved to be most useful for:

- groups of more than 15 students
- finding out what students needed/wanted to know

Table 4.1 Poll activities

Why?	Example	When?	Activity
Finding out what students need	'Questions?' 'What do you need to know?' 'What questions do you have?'	Before session or after first five minutes	Students are asked what they need to know.
Collecting ideas	Free text 'What keywords would you use?' 'What else can you do?'	During session	Students think about a project they have to do, discuss the question in groups and send in their answers. Other groups comment and offer suggestions.
Explaining terms/ clarifying concepts/ aiding recall	Multiple-choice or free text 'What are: • periodicals • journals • articles • leaflets?'	During a session, either before giving the answer or after explaining the terms	Students have to choose the correct term.
Understanding/ remembering/ revision test/ game	Multiple-choice or free text 'No results?' 'Use _____ to search for a phrase' "..." + AND	During a session	Students are told at the beginning of a session that they will be asked questions at various times during the session in order to check that they have been following and understand what is being said.
Engaging students/ checking prior knowledge and assumptions	Guessing game Multiple-choice or free text 'What's the best way to search for a book on the online library catalogue?' • Do a keyword search • Type a keyword and the author's surname • Type the full title of the book and the author's full name	During a session	Students are asked questions at various points during a session, either after being shown a relevant screen or before something is explained. This helps to raise or maintain attention.
Getting students to evaluate a resource, e.g. a database	Voting game Multiple-choice or free text 'Which image database would you use?'	During a session	After using or discussing a variety of databases, students discuss in pairs or groups and vote for their favourite database. If the question is to be answered in free text they can state reasons why they have selected it.
Getting student feedback	Free text 'What did you think of the session?'	Last five minutes or after a session	Used to collect feedback and follow-up questions.

- providing opportunities for individual students to ask questions or voice opinions, and responding to them as they came up
- checking understanding
- providing opportunities for the students to learn and remember new terms
- maintaining interest during a longer session
- getting groups to share and discuss their thoughts and ideas with other groups.

Student feedback indicates that the polls were viewed as being fun and that the students enjoyed using them for various reasons (Figure 4.1). Primarily they liked:

- using familiar devices and systems (their own phones, Twitter)
- contributing anonymously
- seeing their contributions appear as they submitted them
- getting feedback from others.

What did you think of the session?

🌐 **Start** this poll to accept responses

"I thought the session was good. I found out some info I didn't already know, and have a better idea of what to do for my essay, and on referencing. The text in poll made it more interesting. I wouldn't change anything, the session was informative!"
26 days ago

"it was great thanks"
28 days ago

"An enjoyable session with useful info. Loved using the polls."
28 days ago

Figure 4.1 *Sample poll from University of Brighton*

Polls generated discussion within both small groups and larger ones. Even though some polls were embedded in module areas on the VLE, students preferred using them during face-to-face sessions rather than through the VLE before and after sessions. Student feedback showed that a maximum of three to four polls per one-hour session was acceptable. Very few students voiced concerns about the cost of texting, as most of them had free text bundles or used the university Wi-Fi connection and laptops instead. Moreover, a lot of the activities required them to work in pairs or in groups, which allowed them to share the cost of texting.

Summary

Overall, the experience of using Poll Everywhere shows that it is an easy system to set up and use. The most valuable aspect of using this system is that it allows immediate feedback that can be responded to within the confines of the session. Crucially, though, this feedback also allows tutors to respond to individual students, even within a large audience.

References

Deleo, P. A. et al. (2009) Bridging the Information Literacy Gap with Clickers, *The Journal of Academic Librarianship*, **35** (5), 438–44.

Fies, C. and Marshall, J. (2006) Classroom Response Systems: a review of the literature, *Journal of Education and Technology*, **15** (1), 101–9.

Kay, R. H. and LeSage, A. (2009) Examining the Benefits and Challenges of Using Audience Response Systems: a review of the literature, *Computers and Education*, **53** (3), 819–27. doi:10.1016/j.compedu.2009.05.001.

Klaas, J. (2003) Best Practices in Online Polling, *International Review of Research in Open and Distance Learning*, **4** (1), 1–5.

Lantz, M. E. (2010) The Use of 'Clickers' in the Classroom: teaching innovation or merely an amusing novelty? *Computers in Human Behavior*, **26**, 556–61.

Sellar, M. (2011) Poll Everywhere, *Charleston Advisor*, **12** (3), 57–60. doi:10.5260/chara.12.3.57.

Other services

A service that several libraries have experimented with allows users to send class numbers (call numbers) or brief catalogue details from an OPAC straight to their phones via SMS. This sort of 'one-off' service, where you don't need to store users' mobile numbers, is ideal for use with the free e-mail to SMS services provided by most mobile operators. It has therefore been introduced in several libraries (see Figure 4.2) as an early experiment in using mobiles to deliver library services, using in-house expertise to develop the service on top of the standard catalogue.

Send the title and location of an item to your mobile phone

Title: Information literacy meets library 2.0 / edited by Peter Godwin and Jo Parker

Enter your mobile phone #: [   ]

(Use the full 10 digits - no spaces, no dashes. Ex. 6105265000)

Select your provider: [Cingular/AT&T ▼]

Choose an item near you:

1. ◉ S McCabe : Z674.75.I58 I54 2008 (AVAILABLE)

Note: *Carrier charges may apply.*

● Send   ● Cancel

Figure 4.2
Screenshot from Bryn Mawr College's catalogue (http://tripod.brynmawr.edu/)

Summary

Using SMS (or text messages) in most library services is relatively easy to start doing. Text messaging can be done from the most basic mobile phone available and for a minimal cost, so it is likely that most of your library's users have access to it.

It is also likely that initial applications of text messaging within your library service will have minimal costs for introduction. Even the most expensive options – commercial packages that allow large numbers of text messages to be sent – are likely to amount to the cost of one or two hardback books.

As a familiar technology that the vast majority of our users will be able to access at very little cost to themselves or the library service, text messaging is an easy way in to providing mobile-friendly services. Think of it as the 'low-hanging fruit' of mobile services, ripe to be picked before you venture into harder or more expensive mobile options. Try a basic text-a-librarian service or using mobile phones as 'clickers' in information-literacy instruction and build up your mobile services from there!

Note

1 http://en.wikipedia.org/wiki/List_of_SMS_gateways.

Further reading

Brekel, G. (2011) iPads, Mobile Libraries and Medical Apps, *Journal of the European Association for Health Information and Libraries*, **7** (2), 16–18.
    A short, informal article from an innovative Dutch medical librarian. Describes several activities, including loans of iPads, and also a great many interesting mobile apps and websites, plus how the Central Medical Library of the University Medical Center Groningen developed its own web app.
Buczynski, J. (2008) Libraries Begin to Engage Their Menacing Mobile Phone Hordes without Shhhhh!, *Internet Reference Services Quarterly*, **13** (2–3), 261–9.
    A nice summary from 2008 of how libraries were increasingly using mobile technology, particularly SMS, to engage users. Includes links to several projects and services. This paper pulls several text messaging ideas together into one place.
Herman, S. (2007) SMS Reference: keeping up with your clients, *The Electronic Library*, **25** (4), 401–8.
    A relatively early introduction of SMS reference from Australia. This case study shows how a library introduced an unfamiliar service to its staff

and users and provides interesting information on how a new mobile service was received.

Hill, J., Hill, C. and Sherman, D. (2007) Text Messaging in an Academic Library: integrating SMS into digital reference, *The Reference Librarian*, **47** (1), 17–29.
A description of how a US university introduced a text-a-librarian style of service via its e-mail system.

Naismith, L. (2007) Using Text Messaging to Support Administrative Communication in Higher Education, *Active Learning in Higher Education*, **8** (2), 155–71.
Rather than a library study, this is a more generic study of a trial of student contact via text message. The lesson learnt about how students interact with such a service is directly transferable to a library setting.

Luo, L. (2011) Text Reference Service: delivery, characteristics, and best practices, *Reference Services Review*, **39** (3), 482–96.
Looks at different models of text reference service and the reasons for choosing them. Most usefully, it details the factors that a library should consider when launching a text reference service so as to ensure that a high-quality service is developed.

Pasanen, I. (2002) Around the World to Helsinki University of Technology: new library services for mobile users, *Library Hi Tech News*, **19** (5), 25–8.
An early adopter's experience. Irma Pasanen describes how a Finnish library service introduced SMS services in an environment where there was already a high penetration of mobile phones.

Pearce, A., Collard, S. and Whatley, K. (2010) SMS Reference: myths, markers, and modalities, *Reference Services Review*, **38** (2), 250–63.
Have you wondered what sort of questions could be asked using a text-a-librarian service? Are you worried about how fast a response users might expect? This article analyses a large number of text reference queries received during the course of an academic year at a US university library.

5

Apps vs mobile websites

Introduction

Mobile devices are rapidly becoming the main way to access online information. It seems sensible that many of us are investigating ways of making our information more accessible to mobile devices, though often in limited ways.

For most of us, the quantity of information that we provide online has exploded since the early days of the world wide web in the mid-1990s. We still provide some basic information that we may have had online since the start. We show opening times, contact details, news and basic instruction or 'how to' pages and guides. Since then, we may have added library catalogues and resource-discovery tools, chat boxes, RSS feeds of useful resources, social media interactions, videos and podcasts, in addition to the same basic information that we have provided for many years.

Some of this information and the opportunities for interaction may be ideally suited to mobile devices. For example, viewing an online video tutorial on a mobile screen while trying at the same time to apply the skills it is teaching on a fixed computer can be extremely effective. While they are in the vicinity of the library, your users may want to know what times you are open, and so they will find it very useful to be able to check opening-hours information on their mobiles.

However, there is a tension between the alternative ways of offering information to mobile devices. We can do it either on the web, using mobile-friendly formats, or by offering platform-specific applications, or 'apps'. This tension has not yet been resolved. Some of the major issues are covered in this chapter, together with examples of how libraries have addressed the introduction of such services.

Meeting the needs of your users

There are advantages and disadvantages to both methods of providing mobile-friendly online information. Whichever method is chosen, try to make sure that you are meeting your users' needs rather than copying others. No matter how good looking the final result is, unless it provides the information your users want in the format they need, it will be a waste.

A common thread running through all the examples that we can find of best practice in providing mobile-friendly information is that they start by investigating those user needs and desires.

At a minimum, you should find out what sorts of devices your users own. This can be done through surveys, which allow you to ask more than just 'what device' people have. Even a short, simple survey about devices can ask a few additional questions about what sorts of contracts people have with their mobile suppliers (do they pay extra for internet access? do they get an allowance of text messages?) and what sort of device they intend purchasing next (rather than simply what they have now). You can also get a snapshot of the devices currently being used via web analytical software that may already be running behind the scenes of your library web pages. This kind of software can detect the devices being used to access your web pages.

This minimum information, identifying the devices that your users own and use to access your current website, is vital. There is no point in investing time and money to produce an app or website optimized for a platform that few of your users own.

Once you have an idea of the devices that may be used to access any service that you develop, you still need to gather additional information. There is so much that we can offer online and many of us have large numbers of pages on our standard sites. Rather than try to offer everything, we need to collect data on what our users really want. What will they value? How do they want to interact with our information on their mobile devices?

This is where qualitative information comes into its own. Focus groups, allowing several of your users to discuss things together, can be a rich source of information. Let them talk about how they want to interact with the library, where they might expect or want to be able to access information and what sort of information would be valuable to them, rather than about specific services. This can feed into a short list of services and types of information that you can look into developing. You can then move on to running surveys in order to rank these in terms of desirability (and rank them yourself in terms of what is achievable) before deciding on the next stage.

Starting with a short list of detailed services or types of information and asking people directly which ones they want (perhaps through a survey) is risky. Any list may reflect your existing website or your own preferences.

Gathering qualitative information will give you a much richer and truer reflection of users' needs and how you can try to meet them. It is likely to give you unexpected results and will help you to produce something that will be used and appreciated.

Once you have an idea of what devices you need to support, and the services or information that you need to deliver to them, you now have to make the difficult decision: mobile website or app?

Apps

Apps can look great and work brilliantly. Why wouldn't they? When you create one you are designing for one particular platform. You don't need to compromise and make your content suitable for as many different devices as possible. You concentrate on making it work for one particular device, as slickly as possible (Table 5.1). Unfortunately, this is also the biggest drawback – an app works on one platform, and one platform only. Even if you create apps for both iOS and Android devices, you will still be excluding the majority of potential users of your services. Both the iOS and Android operating systems also run on both smartphones and larger-screened tablet computers, meaning that you may also have to consider different apps for smartphones and tablets within the same overall systems.

However, iOS and Android devices account for a large proportion of mobile internet usage within many countries, and this is where your data collection becomes vital. If you know that most of your potential users have iPhones or iPod Touches, then it makes sense to create an iPhone app if you can. If many of your users also have iPads, you may need to consider how an app would display on different-sized screens. The resulting app is likely to have better functionality and to be more appreciated by your users than a mobile-friendly site. If you can see that large numbers of your users have or intend to buy an Android device, then the same goes for that platform. Unless you gather the basic information about devices in the first place, you will be unable to make these informed decisions.

If you've managed to identify a small number of ways in which your users wish to interact with your library via mobile devices, you can build these into an app that will provide real value to your users. Without a doubt, smartphone owners prefer the 'app experience' to browsing the web, and will tend to use a well-designed app in preference to even the best websites.

Apps can be designed to integrate the hardware included in devices with their functionality. These could be serious applications, such as Augmented Reality functions (using the phone's GPS, compass and camera) to show the nearest free computer; or more frivolous ones, like generating book suggestions when the device is shaken. It provides a better, more integrated experience than most mobile websites can.

However, this all comes at a price. To develop an app and make it available in an app store (Figure 5.1) requires significant expertise – and often, licences. Most libraries would need to outsource the production of an app, sometimes to more than one external expert if apps are wanted for multiple platforms and devices. Once you had your app, you would then need to continue to buy in this expertise in order to maintain its working as operating systems were updated, solve issues that users raised and, preferably, build additional functionality over time.

Apple – Apple's App Store contains over 425,000 apps and popularized the idea of selling small apps. Accessible from Apple devices and iTunes, it is the only approved place from where to buy apps for Apple devices.

Android – As suits the open-source nature of Android (produced by Google), there are several places from where to buy Android apps. Android Market (https://market.android.com/) is Google's own store, but alternatives include AppBrain (www.appbrain.com/), Amazon's App store (www.amazon.com/appstore), PocketGear (www.pocketgear.com) and GetJar (http://getjar.com).

Nokia – The official store is Ovi (http://store.ovi.com), but other sites such as PocketGear and GetJar also sell apps for these devices.

Blackberry – The official site for apps is Blackberry's App World (http://us.blackberry.com/apps-software/appworld/), but other sites do have small numbers of Blackberry apps.

Figure 5.1 *Some stores from which apps can be purchased*

The expense of creating an app may mean compromises, so you might produce one that is easy (and relatively cheap) to create, rather than one that meets the key needs of your users (Table 5.1).

There is an additional drawback to apps. The users of your library service need to choose to download an app, as compared to automatically

Table 5.1 *Some advantages and disadvantages of platform-specific apps*

Advantages	Disadvantages
They work on the 'native' platform of a device	Expensive
Can use more of the functionality of a particular device	Difficult to maintain
Can be designed to meet your exact needs	Often need to outsource development and maintenance to external experts
Tend to be well presented on a device	Your users need to choose to install it
Users tend to prefer apps	

experiencing a mobile version of your website when they visit it. This means that you must also invest in promotion of the app in order to gain those downloads. Once the app is installed on a device, you must justify the space it takes up, so it must be repeatedly useful to the owner of that mobile device.

Case study: Seattle Public Library and Boopsie
Lare Mischo
Seattle Public Library, Washington, USA

The Seattle Public Library offered a mobile-specific catalogue in late 2008, based on an existing 'text only' catalogue. This worked even in rudimentary smartphone browsers, but it was crude, at best. We also encountered problems with the quirks of mobile-device browsers, and fundamental usability issues. We had three priority features in our best of all possible worlds: run on pretty well all the major phone platforms; work the same way on all of those platforms; and offer a truly mobile-friendly catalogue. We investigated app development, as well as the time and effort needed to create a mobile-aware website and catalogue. An app would have required either cross-platform development tools (then unavailable at a reasonable cost) or separate development effort for each phone platform. So we ended up favouring a mobile-friendly website, in order to simplify development and make it look (mostly) the same on a variety of platforms. However, with limited resources, neither option was particularly feasible. After some test-bed efforts on both mobile websites and apps, serious development was shelved.

In October 2009, librarians came back from an ALA LITA (American Library Association Library and Information Technology Association) forum with news about Boopsie, a company that was developing an app for OCLC that ran on all major mobile platforms. Boopsie emphasized its experience with the Horizon/HIP Integrated Library System (which we use), its commitment to cross-platform consistency and a clean and quick implementation path. Its app used common mobile interface features across all platforms, and the drill-down style was simple and powerful. The catalogue implementation was specifically designed for mobile use.

Boopsie offers more than a simple, mobile-friendly version of our catalogue. Its 'smart prefix search' means that entering just the first few characters of an author or title word receives real-time feedback as search results are narrowed by each character entered. The searching technology was very attractive, since it mimicked the short and focused interactions of our mobile patrons. But it is also another, and different, catalogue interface. It has the specific mobile-friendly features we wanted but the searching is fundamentally different and a number of catalogue features are not supported – including full reservation facilities, 'my list' private bibliography creation and subject authority searching.

Another feature is the use of Google Docs spreadsheets to create the drill-down menu structure as well as to update all text, images and substantial portions of the look and feel of the interface. Changes made to the Google Docs templates go live within minutes, across all platforms. This ability to modify and deploy the updated interface across multiple platforms, without any vendor interaction, contrasts sharply with the difficulty of customizing most library catalogues. This offers a commercial app with much of the local design control of in-house development, but without the costs.

The Boopsie app supports RSS-based content, in addition to a Google Docs editing method. The Library used this feature to integrate RSS feeds for Library events, podcasts and new book and popular search results into the catalogue app, adding substantial portions of its website. Adding RSS content is as simple as creating a menu element in the appropriate Google Doc and pointing it at the RSS feed. Once the menu structure is set up, no maintenance is needed. As the Library faces continuing budget and staff cuts, this auto-update, hands-off feature is really attractive.

One concern about going to an app was related to supporting the wide variety of platforms, especially new versions of phones. This has not been a problem. Boopsie has a reservoir of phones and testers, as well as expensive tools that mimic multiple phones and platforms. Compatibility problems with new phones have been minor and quickly resolved. Support has been solid.

This app provided a solution that met all of our priorities and added unexpected flexibility and features. In spring 2010 a contract was signed with Boopsie and the app was released in May. We've been very happy with the choice, and it works well for this library.

Alternatives to the standard app development route may be emerging. Companies such as Boopsie (www.boopsie.com/libraries.html) create standard apps for a range of platforms. They populate these standard apps with information that you provide to them in relatively easy standard formats. This standard, 'mass produced' style of app may not look as well designed or sophisticated as one developed especially for your library, or include the exact functionality that your users may be asking for. However, it will be designed to work on the native platforms of most smartphones and will give the 'app' experience to your users.

Mobile websites

If a well-designed app is like a top-end sports car, a mobile website is more like a family run-around. It may not be as good looking, but it is likely to be cheaper, easier to run and accessible to more people (see Table 5.1).

At the most basic level, if your organization has a content management system for its web pages, you may simply be able to turn on a mobile style sheet that is present behind the scenes but currently unused. This will detect mobile devices visiting your site and reformat the pages to be more accessible to devices with small screens. There are also many mobile style sheets available online that you can adapt for most websites, making the sites automatically more accessible to mobile devices.

If you have a website and someone capable of maintaining it, then you already have the basic skills in your organization to make your web pages more mobile friendly. The highly specialist skill sets and licences necessary to build apps are simply not needed for a basic mobile site.

To create a truly mobile website, however, takes more time and effort than simply applying a mobile style sheet. A good mobile site should be simpler and flatter than most standard library websites, allowing quick access to the functions or information that people need most. A mobile website should therefore address a small number of areas well – those areas that you have already identified through talking to your users. Don't try to duplicate everything on your website, because your mobile site simply will not work.

Mobile sites are unlikely to look as sophisticated or as slick as a good app but, unlike apps, they should be accessible from the vast majority of smartphones. They can often echo the app experience, looking similar in style on the web page to an app, but because they are built to suit all smartphones they will never look quite the same as an app designed for a single platform and size of screen (see Table 5.2).

You don't need to promote a mobile version of your normal website. Someone who visits your website from a mobile device should automatically be diverted to the mobile version (unless they choose otherwise).

Table 5.2 *Some advantages and disadvantages of mobile websites as a platform for content delivery*

Advantages	Disadvantages
Cheap and easy to produce a basic mobile site	Don't take advantage of functionality built into specific devices
Easy to maintain	Typically less well presented than an app
Work with all smartphones	

Case study: The Ryerson Mobile project
Graham McCarthy
Innovative Technologies Librarian at Ryerson University and Project Manager for Ryerson Mobile

Introduction

Ryerson University is a mid-sized urban campus located in the multi-cultural heart of Toronto, Canada. It is known for its innovative programmes, built on the integration of theoretical and applied learning. Approximately 30,000 students are enrolled in 100 programmes that are distinguished by a professionally focused curriculum and a strong emphasis on excellence in teaching, research and creative activities. Ryerson is also a leader in adult learning, with the largest university-based continuing education school in Canada.

The Library and Archives house a collection that includes over 600,000 books, 1,700 hard-copy serials and an impressive electronic resources collection of over 33,000 full-text journals. The Library has many innovative and supportive services, including a strong reference and instruction presence that is increasingly mobile and virtual. A rapidly expanded research mandate complements the university's traditional undergraduate focus with new frontiers for emerging digital library support. The Library has a staff of over 85, including 25 librarians.

Ryerson Mobile: the beginnings

Ryerson Mobile (http://m.ryerson.ca) is an on-going project that was launched on 14 September 2009. Planning for the project began early in February 2009, when the initial applications were decided on and early system schematics were created.

It was also decided that students should play a major part in the creation and delivery of the mobile site, as we wanted it to be as useful as possible for the students. We hired two fourth-year Image Arts students and one fourth-year Computer Science student to work on various components of the system. Two librarians oversaw the project on a daily basis and formed partnerships with other collaborating departments that were involved in varied capacities for some of the apps development.

Collaboration

The project has been a true collaboration, involving the efforts of many campus groups and departments. The Library and the University's central IT unit (Computing and Communications Services) manage the project and contribute to its core development and on-going maintenance. The Office of the President has played an important advisory role in the direction and focus of the project, as well as in publicity and marketing. The School of Management, specifically the Students in Free Enterprise (SIFE Ryerson) group, performed the early stakeholders' analysis, user focus groups and surveys and created the initial marketing packages. The Computer Science department has involved its students in the project by holding class projects to improve parts of the application's functionality, such as improving the campus navigation system through geo-location. The Ryerson Mobile project has been a true reflection of this University's integrated and multidisciplinary operation in creating an immersive campus for the students.

Apps

The current application suite for Ryerson Mobile is listed below (Figure 5.2).

- **Campus Directory:** allows users to look up Ryerson staff and faculty by name, phone number, e-mail address and office location.
- **Class Schedule:** lets students view class information for the current and next semesters. Schedules can be viewed daily, weekly or by course code.
- **Campus Map:** shows the location of Ryerson campus buildings and can draw a path to a desired destination, based on the user's geo-location.
- **Book a Room:** lets students search for and reserve study rooms in the Library.
- **Library:** provides users with the facility to see open hours, perform mobile research, search the catalogue, make reservations, renew material and check fines.
- **Find a Computer:** shows the real-time status of computer availability in the Library and across the campus. It also provides information on the availability of the laptops that students are able to borrow from the Library.
- **News:** provides access to the top news stories coming from various news sources on campus.
- **Profile Configuration:** provides personalization options for users to customize their experience.
- **OneCard:** allows students to view food credit for canteens and print card balances for IT facilities as well as locate machines that can re-fill money onto their cards.
- **SoapBox:** an extension of Ryerson's Idea Management System that allows students on the go to add ideas for improving the campus and to vote on ideas submitted by their peers.
- **Admission:** provides prospective students a place where they can ask questions and view information pertinent to the University.
- **TTC:** a link to the Toronto Transit Commission's mobile transit application, which provides bus, train and streetcar schedules to the students.

Figure 5.2 *Ryerson's mobile web app*

Mobile web app

A decision was made early in the life of the project that Ryerson Mobile would be a mobile web app rather than a device-specific native application. Among the reasons for this decision were that maintaining multiple code bases and developing expertise on various platforms might be too much work for a light-weight project and the costs associated with licensing and development kits could be very costly. Tools such as Appcelerator and Phone-Gap are great examples of one web code base being utilized to generate native mobile applications; however, at the time, these tools were not as powerful as they are now.

Our mobile website was designed to be accessible by any mobile device with internet access. We maintain three separate views of the same data: a high-graphic version for newer smartphones that support WebKit (i.e. iOS, Android and BlackBerryOS 6+); a medium-graphic version for the older smartphones that support graphics and JavaScript but not WebKit (i.e. BlackBerry OS pre 6); and a basic WAP version to cover the rest of the devices that don't support JavaScript or images. This device-agnostic approach allowed our developers to focus on one platform for delivering all of the services.

To take advantage of device-specific features, such as the camera and local storage for storing login credentials, we created native applications for Android and iOS devices that operate as a wrapper around the web application. We did not replicate any of the web apps' functionality, but created a native application bar at the bottom of the page to provide these additional features.

Conclusion

The Library has traditionally been an information steward with a strongly service-oriented ethic; however, it has not always had the technical capacity to create truly innovative services. Forming relationships with other campus partners has enabled the Library to meet its service goals and to contribute its strengths to broader campus initiatives such as Ryerson Mobile.

Up to now, we have suggested that a mobile-friendly website is normally quite basic. This may be so, but it definitely isn't always the case. You can create incredibly sophisticated mobile sites that work across a range of platforms. In the UK, Mobile Oxford (http://m.ox.ac.uk) contains a wealth of information for the university and city of Oxford, displayed in an app-style format. This includes information from the confusion of different libraries under the umbrella of the University of Oxford. You can search for a book, look on an interactive map to find which libraries stock that book, discover details of the libraries and even plot routes to get to them. It is a fantastic example of a well-executed mobile site that includes a range of functionality.

Summary

Before creating a mobile website or app, first consider your users. What sorts of devices do they have or are they likely to own in the near future? What devices are being used to access your current web pages? What are the key information or services on your web pages that users are likely to want to access from mobile devices? Once you have this information, you need to consider your own situation. Can you afford to commission and maintain apps for the devices that your users own? What skills do you have in-house to develop mobile-friendly web pages or apps? If you have the skills available, can the people with those skills dedicate time to developing services, or would you need to outsource development?

You can then pull the answers to these questions together and decide whether it is worth developing apps that deliver a great user experience, whether a mobile version of your site may be more realistic, or whether you can have a combination of the two, with a narrow range of information and services available through an app and additional information accessible through mobile-friendly web pages.

Every library service is likely to end up with something slightly different, but doing nothing is an increasingly unacceptable answer. Our users expect to be able to access our online information via their mobile devices, and so we need to decide how we will deliver the information that they need to their mobile devices – not whether we will do something about the issue or ignore it.

Further reading

Braun, L. (2010) The Big App: New York's libraries take homework help mobile – with a little help from their friends, *School Library Journal*, **56** (12), 49–51.

A suite of apps has been developed by New York Libraries to help school children to do their homework. The key lesson that it found is applicable to everyone – listen to what your users want. By involving children (the target audience) in the project from its early stages, it created a much more useful resource than it could have done otherwise.

Cutshall, T., Blake, L. and Bandy, S. (2011) Creating a Mobile Library Website, *Computers in Libraries*, **31** (7), 23–5, 48.

This article outlines a university's experiences of creating its own mobile library website. Written from a practical point of view, it is a useful insight into the process.

Murray, L. (2011) The Best Things in Life Are Free (or Pretty Cheap): three mobile initiatives that can be done now. In: Woodsworth, A. (ed.) *Librarianship in Times of Crisis* (Advances in Librarianship, Volume 34), Emerald Group Publishing Limited, 139–57.

In times when many of us are seeing funding restrictions, this seems an appropriate book chapter! It exhorts us to do three things that need cost very little, beyond staff time. SMS reference is one, but the other two relate to creating a mobile web presence and a mobile-friendly catalogue. If you need more reasons for doing such a thing, this chapter may encourage you!

Power, M. (2011) *Mobile Web Apps: a briefing paper.* JISC Centre for Educational Technology and Interoperability Standards, http://wiki.cetis.ac.uk/images/7/76/Mobile_Web_Apps.pdf.

This briefing paper gives a great overview of, and argument for, the use of mobile web apps to deliver services in educational institutions. It argues that, unless you need to access the full hardware capabilities of a particular mobile device (for example, for a high-end game), mobile web pages, presented as web apps, can deliver a high-quality service that overcomes the need to develop for a range of closed ecosystems (Apple, Android, Microsoft, Blackberry etc.).

Rapp, D. (2011) Boopsie Rolls Out Mobile Checkout Feature, *Library Journal*, www.libraryjournal.com/lj/home/890236-264/ boopsie_rolls_out_mobile_checkout.html.csp.

An interesting development, though perhaps quite limited, due to security issues. This brief article in the *Library Journal* describes how Boopsie (which produces apps for libraries) has started to build into its apps the ability to issue books for some libraries. In the article, this functionality was used at Cuyahoga County Public Library. Since Boopsie produces largely generic apps for many libraries it could easily be rolled out to a large number of libraries. It can be used only in libraries without security gates, as the app cannot remove security features, and so has limited use at present.

Wilson, S. and McCarthy, G. (2010) The Mobile University: from the library to the campus, *Reference Services Review*, **38** (2), 214–32.

Ryerson University has done significant work in mobile delivery of library services, including developing a mobile site and an app. This article reviews the library's experiences in creating mobile services, and points in particular to the opportunities for partnership and bridge-building between the library and other groups around the university.

Wong, S. (2012) Which Platform Do Our Users Prefer: website or mobile app? *Reference Services Review*, **40** (1), 103–15.

A study at Hong Kong Baptist University Library, where videos were offered through a choice of app or website. In this case a roughly equal split was found between use of the website and the app. The reasons why are discussed.

6

Linking physical and virtual worlds via mobile devices

Introduction

We hear occasionally of schools and colleges that have been built with no library. Public libraries are being closed because it is felt that they are there just to loan books and that fewer people need this service. Funding to many library services is being cut because of perceptions that learners can get everything they need from the world wide web.

Some of this could be because our physical libraries are getting increasingly divorced from the rich variety of online as well as physical resources that they hold. They are repositories for many physical items, such as books and print journals. They also often serve as study space, with desks, computers, and perhaps rooms for activities such as group or silent individual study. Physical libraries are also a place for library and information professionals and para-professionals to base themselves and offer help and instruction.

But these functions no longer necessarily sit together very well. You don't *need* to have physical study space co-located with physical copies of books if the members of your library use most of their information resources online. If the majority of your information resources are easily accessible online from anywhere, by focusing library help within a physical location you can miss many of the people who may need help from the library's professional staff.

Even the traditional print resources often don't meet our users' expectations. People who are used to searching for information using Google, YouTube or Flickr can find a physical library confusing and difficult to navigate. We send people to clunky-looking catalogues that are available either online (but rarely mobile friendly) or in fixed locations within our buildings. These catalogues may then provide minimal information about items, including a secret librarian's code (call number) that users must attempt to decode in order to find the item. Only when (or if!) a user physically finds the item can they really start to assess whether it will be useful.

These are just the physical items, which are often the most visible reflection of the information that libraries hold. Electronic resources increasingly make up the bulk of many libraries' collections, and by their very nature can be more concealed, or seen as being less directly linked to the library that provides them. Normally the only clues to these electronic resources within the physical library space are printed hand-outs on a circulation desk, or the odd poster extolling the virtues of a particular database.

Mobile devices can help us to start bringing the physical and virtual worlds together. They can bring electronic resources into our physical spaces, or bring physical items to life within the electronic world. They are fantastic bridging devices, if we choose to use them, making our physical spaces alive with the richness of our collections and expertise. Consider how much more visible we can make our resources if we can blur the boundaries between the physical and electronic worlds within our libraries. Mobile devices can start to make this happen.

This chapter outlines a few different types of technology that can help us to achieve all these aims: QR (Quick Response) codes; Near Field Communications (NFC); and Augmented Reality (AR).

QR codes

QR codes are two-dimensional matrix codes. The traditional barcodes that most people are familiar with encode information in only one dimension (the height of the lines contains no information). Matrix codes, such as QR codes, encode information both horizontally and vertically. This means that they can encode more information in a limited space than a traditional barcode can.

There are many different types of matrix codes, but the type covered in this section is the QR code. Other competing codes may become more popular in future, but QR codes are the ones nearest to the mainstream. They are usable by the largest numbers of people, and accessible through a range of (normally free) applications (Figure 6.1).

QR codes originated in Japan in the 1990s and were originally designed by the company Denso-Wave to help in tracking car parts. I myself have worked in production-planning roles within manufacturing companies, and each stack of parts was moved around the factory with slips of paper containing information about the parts, the processes they needed to go through, where and when they were needed and more. A single matrix code attached to each batch could have readily contained all that information, but a barcode never could. It could also, of course, have been scanned at high speed by a small

Kaywa – http://qrcode.kaywa.com/
Generates codes in several different sizes to encode URLs, text, phone numbers or SMS messages.
Zxing – http://zxing.appspot.com/generator/
Generates codes in several different sizes to encode URLs, text, phone numbers, SMS messages, contact details, calendar events, e-mail addresses or geo-locations.
I-nigma – www.i-nigma.com/CreateBarcodes.html
Generates codes in several different sizes to encode URLs, text or SMS messages. Also includes the option to easily include a title that appears in plain text underneath the QR code.
BeQRious – www.beqrious.com/qrcode/create
Generates codes in a set size to encode URLs, text, phone numbers, SMS messages, e-mail addresses, contact details or geo-locations. Also has additional options, including the ability to easily insert simple images inside the QR code.

Figure 6.1 *Some websites that can be used to generate QR codes*

portable scanner and all the required information read and updated quickly and painlessly.

Though Denso-Wave created the standards for QR codes and owns the patent on them, it has chosen to make the standards freely available for everyone to use. As a result, many people have created free software to create and read QR codes. They were designed in 1994 to be quick and easy to scan and translate with scanners, so the typical camera phone available today can cope with them readily. This combination of hardware capability and freely available software means that the majority of mobile phone users are able to use QR codes – if they know what they are and if they choose to download a freely available QR code scanner to their device. These are two very big 'ifs', however, meaning that QR codes are still little used in most environments.

QR codes are heavily used in Japan, perhaps due to QR code scanners' having been included on many phones by default for the last several years, meaning that users didn't need to choose to download an application themselves. With scanners available on most mobile phones, manufacturers and others took advantage of the Japanese people's love of new technologies and QR codes quickly started to appear 'in the wild'. They now seem to be everywhere, linking to everything from product information or offers on supermarket shelves, to movie trailers accessible via codes printed on cinema posters. With large parts of the population owning a QR code scanner (a mobile phone with a ready-installed application) and being aware of the codes, and the codes' increasing incidence, a momentum was created.

That degree of momentum hasn't yet been achieved in most countries. Occasionally mobile operators (such as Nokia) have experimented with including QR code scanners on certain phone models in particular countries, but this has been limited. It may be about to change, however, with barcode scanners (which can read QR codes) being included in some default Android installations, and with increasing numbers of apps for all smartphones also being able to read the codes. Users can download geo-caching applications (linking physical objects with GPS locating technology), scanners to compare prices between shops, or games that include QR scanners as part of their functionality (Figure 6.2). Thus, increasing numbers of users have QR codes scanners that they can use, even if they haven't directly chosen to seek them out.

Although URLs are provided here, most of the applications should be searched for in an appropriate app store.

Kaywa – http://reader.kaywa.com/
   A good general QR code scanner that works with most camera phones.
I-nigma – www.i-nigma.com/Downloadi-nigmaReader.html
   Another good general QR code scanner that works with most camera phones.
Stickybits – www.stickybits.com
   Can scan barcodes and many types of matrix codes such as QR codes. Information about those items can be stored in stickybits, allowing competitions, online comments and other metadata to be associated with anything that has been allocated a code.
SCVNGR – www.scvngr.com
   This is a scavenger-hunt type of application. One of the ways used to prove that you have been to a location is by scanning a QR code at that location. This is increasingly being used within gaming applications to prevent cheating when you are checking in!

Figure 6.2 *Some apps that can be used to read QR codes*

Alongside this, QR codes are increasingly appearing on products and on promotional items, driving up awareness amongst the population. In the UK, where I am based, I've recently seen them on drinks bottles (linking to a mobile site), movie posters (linking to trailers), in DIY stores (linking to 'how to' videos), flyers (linking to competitions) and, most recently, even on a student's research questionnaire (linking to an online version of the questionnaire).

The increased availability of QR code scanners on mobile phones (meaning that you don't need to download one), together with the increased awareness engendered by their appearing in increasing numbers of locations, means that we may well be on the verge of their mainstream, popular adoption. Within the next couple of years many countries may reach this tipping-point, whereby they will become as ubiquitous as in Japan.

So what? What does mainstream adoption of QR codes mean for libraries?

How can we use them to start bridging the gap between the physical and virtual worlds of our libraries?

A few ideas are listed below, but across all these ideas one must also exercise an element of caution. In whatever way you choose to use them, it must add value to your service. This will be different for each of us, so don't try to copy every idea that you see. Instead, think about the applications that will make the biggest difference to your users. Until QR codes become completely mainstream there will be perceived barriers to using them and they must offer a genuine benefit to your users, in order to overcome those barriers.

Linking to phone numbers for support

Realistically, most devices used to read QR codes will be mobile phones. So why not use them to make phone calls and send text messages for context-appropriate support? QR codes can prompt a mobile phone to ring a number or start composing a text message. You can therefore use them to link to appropriate numbers from physical (or virtual!) locations. A QR code next to a printer or photocopier might prompt a phone call to an IT desk. All of your IT equipment could carry a sticker with a 'report a problem' QR code that prompts a text message containing equipment details to an IT support number. A QR code next to a library catalogue might text or ring a subject specialist for help in finding items. When items are issued from your library you could even automatically print QR codes on the receipts which, when scanned, could send a text message asking for those items to be renewed.

Case study: QR codes at George Fox University Libraries
Robin Ashford
George Fox University, Oregon, USA

Introduction
The exponential growth of internet-enabled mobile devices led librarians at George Fox University (GFU) to move forward in implementing QR codes in their libraries. The goal was to provide easy information access and value-added content to our mobile users.

Background
George Fox is a private university with just under 3,500 students. The main campus and library is in Newberg, Oregon, USA. Three satellite centres serve primarily graduate and

doctoral students. The largest of the centres, the Portland Center in Portland, Oregon, serves approximately 1,100 students and also includes a physical library. The librarians and staff at both libraries are service oriented and work hard to provide quality resources and services.

Two of us had been closely monitoring the mobile movement. After learning about QR codes we began experimenting with the technology as a way to link print to digital content. Finding the QR codes easy to use and feeling that the benefits gained could be substantial, we shared our interest and ideas for implementation with the Library director and other librarians. Soon afterwards, we began to implement QR codes in our libraries.

With a total of six librarians between the two libraries, formal committees are not the norm at the GFU libraries. A one-hour librarians' meeting held weekly is the primary avenue for bringing up new ideas and receiving feedback. The two librarians from the Portland Center drive in to attend these weekly meetings on the main campus. Librarians often take the initiative on minor issues and will move forward either on their own or with others who may be interested.

On the main campus the librarian who had been experimenting with QR codes recruited another librarian and a staff member to assist her in implementing QR codes in the main library. For her main project, the codes were batch created by the technical services librarian for the popular magazines-reading area. The library staff member and a student worker created a QR code instruction guide that was placed on the shelf ends of the magazine area and taped the QR codes in place following a consistent pattern. When they scanned the codes, mobile users would learn that the print magazines and periodicals were also available online. The QR code was a link to a search for each journal in our A–Z list, showing the full-text date range and a link to the database. This was not a mobile-formatted page, so although users could follow the links to the content, it was not likely that this would happen. It was felt that placing the QR codes there for users to learn that the content was available online was still worthwhile, even if they didn't actually use their mobile devices to access the content.

Another librarian from the main campus came up with the idea of placing a QR code on the door to our main study room that would take Library patrons to a room-reservation page on the Library website. Again, this was not a mobile page, and although the form could none the less be filled out on a mobile device, it was a lot of work. If the form had required less information and had shown a calendar of availability, it would have been a more compelling implementation.

At the Portland Center Library I experimented, and researched ways in which QR codes were being used successfully in retail marketing as well as in libraries and higher education in general. My conclusion was that QR codes that linked to quality videos would be appealing to many. Short videos work really well on handheld mobile devices and convey a lot of information. That also meant that we could create QR codes for the professional video trailers for our educational DVDs and audio books (for those that had trailers) and for librarian-created video content (short tutorials, in our case) that was hosted on YouTube. I made individual QR codes for a few videos and audio books in our collection and fixed them to the fronts of the CD/DVD cases. A short instructional sheet about QR codes was placed on the ends of the stacks where the videos and audio books were housed.

Another implementation has been for art shows. Our art display changes every month or two, and when an artist has a website I create a QR code for the site and place it on or near the artist's statement. This allows those who are interested in doing so to learn more about the artist and to view additional works, thereby adding value at the time of need or interest.

Finally, in order to publicize our recently acquired mobile catalogue, we involved the university marketing department and sought input from all library staff at both libraries. The mobile library site and catalogue interface was provided by a third party (for an annual subscription) and we wanted to let people know about it. Our marketing department's graphic designer created a card – larger than a business card, but smaller than an index card – to promote our new mobile site. Two designs were drawn up, both featuring a QR code in the middle, along with the GFU logo and a simple text stating 'New Library Mobile Website – Scan the code or go to libraryanywhere.com/m/282.' Library staff voted for their preference, and cards were produced and placed around campus. We also made them available during library instruction sessions. One of our librarians also made an informational page for our non-mobile website explaining QR codes to users not familiar with the technology.

Lessons learned

Many are still unaware or only vaguely aware of QR codes; therefore education is critical. Even after learning about QR codes, users may lack sufficient motivation to take the time to download an app and scan a QR code.

Creating QR codes that help to solve a problem or that provide related information on mobile-friendly sites at the time of need/interest are the most useful. QR codes that add little value should probably be avoided, as they may dissuade users from using QR codes in the future (just because you can, it doesn't mean you should).

No reliable assessment mechanism was in place when we began implementing QR codes and, without any data, it's difficult to know how much scanning has taken place. The general consensus is that the numbers are low at both libraries; statistical data would have helped us to learn which codes had been most and least used.

Conclusions

QR codes can provide an easy and useful way to deliver information or offer a service (e.g. scan a code to ask questions in live chat with a librarian). As a low-threshold (low cost, easy to use) technology there is plenty of potential for their use in libraries, at least until the next newer/better mobile technology comes along.

Thoughtful planning, and taking time to strategize ways in which to address the lack of education/motivation, could encourage the use of QR codes and further benefit our mobile users. Interviews about users' attitudes toward QR codes could provide additional insight.

Gathering statistical data on QR code usage for assessment (via a number of available analytical services) would be most important for larger institutions or major marketing campaigns. Smaller libraries could also gain a better understanding of usage from statistical data, and this would be helpful for future planning.

Linking print items to their electronic equivalent

I always feel sorry for people who are standing in the journals stacks, searching through copies of print journals, or standing and looking at bare shelves where the multiple copies of a core text-book stood before their colleagues beat them to the stock. To save these frustrated users, you could use QR codes to link to electronic versions of popular texts from the appropriate physical shelf location, making discovery of these items easier. For the most popular books this could be a 'fake' book, a cardboard box or insert containing a copy of the book cover on one side and a QR code linking to the electronic version on the other. Alternatively, you could place posters on shelf ends displaying QR codes that link to electronic versions of books on those shelves, or one generic QR code that links to a search box for your main e-book supplier, allowing library users who are browsing the shelves to find electronic items while they are searching physically for print books.

You could also link to electronic journal holdings by placing QR codes next to copies of current journals, or next to boxes of past copies. At the University of Huddersfield we printed stickers with automatically generated QR codes, each of which is linked to the appropriate ISSN of a journal within our link resolver and which, when scanned, always brought current information for our electronic holdings. A more sophisticated version of this approach might be to link to an online search tool and predefine a search for a journal identified in the QR code, allowing the user to then refine the search with their own particular subject. This could satisfy the desire of some users to browse their 'key' journals, at the same time inserting the convenience of online subject searching into the process.

A similar idea could be applied to an electronic table-of-contents service, bringing up recent tables of contents for the item scanned. Users could save the page to their phones, which of course would be automatically updated as each new edition of a journal was released. Tables of contents tend to include RSS feeds as standard, which can feed into RSS readers that are available for all smartphones.

Enriching print hand-outs and help sheets

This is the main way in which I currently use QR codes. We often produce hand-outs for classes taught by library staff, as well as generic hand-outs that are available to everyone. We also produce online help materials and videos. QR codes are an easy way to bring these two sets of materials together.

On printed hand-outs, I regularly embed QR codes linking to videos on YouTube – sometimes produced by us, sometimes by third parties. I link to useful websites on the material, including tutorials on suppliers' websites. I sometimes even embed mobile-friendly quizzes to help people check their knowledge. These sorts of ideas turn flat, static hand-outs into interactive ones, extending their usefulness. Because not everyone uses QR codes, the information is always additional to the basics contained in print, but the codes can really add value for those people who wish to interact with them.

For this idea to be practical, you need to bear mobile usage in mind for all your information-skills online materials. If by default you create materials that will be mobile friendly, it is easy to create a QR code to link to them. If they are in formats that do not work well on mobile devices, then simply including a QR code linking to them from a hand-out will not work. Try to make it easy for yourself by creating materials that work in many formats from the start, allowing you to easily reuse them in ways such as those outlined here.

Linking from posters

We often use posters in libraries to advertise online services. Wouldn't it be great if we could link from the pieces of paper on the wall directly to the services? QR codes are one way of doing just that. A small QR code linking to the appropriate URL can be an easy way of taking users directly to that database or library web page, or a booking form for that library class or group study room advertised on the poster.

If you carry out library surveys, you could link directly to the survey from a poster, or straight to a competition page that you might be running. My children do a summer reading challenge every year at the public library: they could scan a QR code themselves at the library each week to qualify for the award at the end of summer, instead of relying on staff to check how often they have visited.

Embedding help into physical objects

Think about all the questions that might be asked in your library about using equipment. Wouldn't it be nice to have 'how to' videos embedded into the equipment itself, so that at least some people could find out for themselves how to carry out a task rather than asking the staff? QR codes could help with this.

If you can identify commonly asked questions about equipment and then create short videos to show the answers, these will be far more useful than printed guides. It is much easier to show someone (such as via video) than to explain using text, or even images, on a help sheet or poster. The videos could be hosted for free on a service such as YouTube, which also delivers them in formats appropriate to the devices used. You could create videos for all sorts of equipment, such as printers and photocopiers within the library, and also for any loanable equipment, such as camcorders or audio recorders, that you may have.

An alternative to QR codes

There are many alternative formats available that are similar to QR codes. One that is a little different is a proprietary system that has been developed for Sony's PSP.

Sold particularly to educational institutions, 'Second Sight' from connectED (www.connectededucation.com) translates proprietary codes on the fly in order to generate an Augmented Reality experience on the PSP.

Another great physical location in libraries where you can embed help is on shelf-ends. Placing QR-code links around the library to context-appropriate help need not take up much space on a shelf-end. This is space that most of us have available, so consider using QR codes on shelf-ends to link to subject-appropriate tutorials, or to catalogue searches for material recently added in that location.

Storing information for later use

In my library, in recent years we've noticed increasing numbers of students taking pictures of the library catalogue screen rather than writing down a book's details. We also have them arrive regularly at our information points with brief details of items noted down on their mobile phones. Why not take this one step further and use QR codes to allow them easily to take the results of catalogue searches with them?

You can generate QR codes live on your catalogue, linking either to text containing the key information of a record or to a mobile-friendly version of the same catalogue record. This would allow users to scan the codes of useful results and take the record with them, rather than relying on a poor-quality photo or incomplete notes that they've taken down.

This idea could also apply to things such as printed guides. It is easier than ever to transfer files between fixed and mobile devices. Why not offer QR codes linking to PDF versions of your print guides? People could then carry these with them and have them easily to hand when they need them, whether directly on their mobiles or transferred across to larger-screened, fixed computers, or even to their e-book readers. This transferring of information is increasingly easy, especially with the increase in popularity of online storage services such as Dropbox (www.dropbox.com, see page 35), which allow you to access files from a range of devices. Offering a single poster of QR codes that link to your main guides, or appropriate QR codes at locations where print guides are left for users to help themselves, could improve your users' perception of your service. It may even help to reduce your library's printing costs if fewer people take your printed guides and instead can easily access the same content via all of their computing devices.

Linking to other materials from relevant places

There are lots of other types of materials that you could link to from relevant places by using QR codes. Think of the kinds of services, particularly online services, that you produce and where it might be useful for members of your library to have that information available. Health librarians who produce current awareness alerts could link to a page containing their latest bulletin via a QR code, perhaps near the books or journals that their specialists are likely to use.

On your web pages, where people find links to log on to your online resources, why not include a QR code linking to a tutorial for a resource next to the link to the resource itself? In that way it would be easy to have the resource open on a fixed PC while working through a tutorial or video on a mobile device.

Again, in a similar place online, or near to appropriate study spaces, you could link to a timetable of classes that you offer, with the option to download it to a calendar on the user's phone.

Summary of QR codes

QR codes are free to create and use, can be used by nearly everyone and almost seem like magic the first time you use them. There are vast numbers of uses to which they can be put in libraries, only a few of which have been suggested above.

However, not everyone is willing to download an app in order to start using them, not everyone who has an app will use it and users really do need a good reason to scan codes. So, make sure that you focus on creating QR codes that are directly and transparently useful. Don't try too many things, but instead try something that users can readily see will be of benefit to them. This will be different for each of us, so think about your users' needs rather than about creating QR codes just so as to jump on the bandwagon. If you find the right uses, then users will love them!

The biggest mistake you can make with QR codes is to try to force people to use them. You can spend large amounts of time and effort promoting them, to very little benefit. If you find the right uses for your library and your users, then you need to do relatively little promotion or training and they remain a low-cost, easy innovation to introduce. QR codes should only really need an initial push so as to make people aware that you are using them, rather than serious, on-going promotion and training.

Case study: University of Bedfordshire mobile developments
Peter Godwin
University of Bedfordshire, UK

Libraries have moved on from the hybrid library, are facing a public perception of the electronic delivery of all content, and Web 2.0 has given everyone the potential to author and share information. Now add to this the phenomenal spread of mobile devices around the world. Information is becoming mobile and social. It also appears that the recession that has hit many parts of the world may not slow these developments. Over coming months librarians will be spending more time conditioning their services to the mobile environment, and the University of Bedfordshire is no exception.

The University of Bedfordshire (UoB) is a new university 30 miles north of London, with libraries at two major campuses (Luton and Bedford) and several smaller units, serving the 23,000 students. International recruitment is particularly strong and I shall focus on the views of these students later. A small general survey in summer 2010, together with the obvious prevalence of mobile devices everywhere among the students, prompted the development of mobile-friendly access to our resources. Mobile devices also seemed to have potential for connecting with our users to aid information literacy development.

In the past year we have developed an in-house free learning resources application for Android phones that links to the catalogue, our website, a list of our 'Just a Minute' videos with ability to view, a GPS 'Find a Campus' guide to the local campus, and e-mail through the auto-enquiry function or ability to call us directly to renew items (see http://lrweb.beds.ac.uk/libraryservices/whoweare/apps/android).

We then needed a direct library catalogue application and for this we chose the commercial LibraryAnywhere software, which is compatible with iPhone, Android and Blackberry devices (see http://lrweb.beds.ac.uk/libraryservices/whoweare/apps).

Finally, we have chosen the commercial Z-Bar software for both iPhone and Android devices, which allows the scanning of barcodes (importantly, of book barcodes in shops) to match against our library catalogue (see http://lrweb.beds.ac.uk/libraryservices/whoweare/apps/barcode).

Although this seemed such a good idea, its usefulness is greatly diminished because our branch of Waterstones bookshop in Luton has since been axed!

We wanted to experiment with using mobiles to exploit our stock. QR codes offered the most promising way of helping with this and also to help explain procedures or how to use baffling equipment on site. We had already developed a series of short library videos which we called 'Just a Minute'. There are now 26 of these covering anything from basic introductions to the catalogue to print credit machines, self-issuing of books and referencing (see http://lrweb.beds.ac.uk/libraryservices/whoweare/videos).

QR codes offered a novel and exciting ability to put links to these at points of use; for example, a video on how to operate a new movable stack to house our journal collection at the Luton site. We have experimented with the use of QR codes in subject leaflets in order to encourage connection with our best resources. Certain text-books have always been in exceptional demand. Although these might sometimes be available as an electronic book, our users still opted for a print book. We have therefore been using QR codes on the shelf-ends to promote the e-versions of our most popular titles (Figure 6.3).

Figure 6.3 *Promotional material at the University of Bedfordshire*

So far I have listed our innovations without any indication of their impact. Downloading of the mobile applications has been encouraging. Take-up of QR codes has been slow, and was likely to be so, due both to students' being unaware of their purpose and to the barrier of having to download a free application like Beetag or I-nigma to their device. The need to log in to the wireless network on campus provided another time barrier. Therefore promotion was required, and championship by academic librarians, in order to encourage their use by the students.

In April 2011 I decided to take a snapshot and test the ownership of devices, and knowledge and take-up of our mobile services, with my major client group: large cohorts of Business Masters students. The vast majority of these students were from the Indian subcontinent. I managed to gather responses from 172 students (about 10% of this student cohort), of whom 116 considered that they were using a mobile device as a learning tool, but only 25 admitted to having an iPhone, 20 an Android, 3 an iPad, 2 an e-book reader and 24 a Blackberry. Only 27 said that they knew of our Android application for library services and 10 had used it. Only 36 knew of QR codes, with 21 using them. Many of these students had seen a demonstration of a QR code at an earlier date. Despite some deficiencies in the survey design, the message was that these students were not so advanced in mobile use as I might have supposed. Three follow-up focus groups in June 2011 with nine students on the same course revealed a similar lack of awareness and take-up of mobile facilities. Familiarity with and use of these devices will be crucial in a business environment, so I will need to be much more proactive with cohorts this academic year. Mobile devices offer an excellent way to keep in contact with large groups of students and drip-feed them with information at the right time.

The prime message from this case study is that it is still early days in the use of mobile devices for delivery of our services, and particularly for information literacy. This will undoubtedly change. QR codes are becoming more mainstream – I saw one in our St Albans branch of Waterstones only last week!

Radio Frequency Identification and Near Field Communications

Near Field Communication using Radio Frequency Identification (RFID) chips has been slowly penetrating our lives for years. Within libraries, RFID allows the implementation of easily used self-issue and -return of stock, and this is the use through which you may be most aware of the technology. Each item of stock is tagged with a small sticker containing an RFID chip and antenna. The self-service machines associated with the technology contain a small radio transmitter and receiver to communicate with these chips.

Libraries also use RFID in users' library cards. This allows identification when the card is close to a sensor, rather than needing to swipe a magnetic strip or scan a barcode.

However, RFID technology appears in many other places besides the book stock of a few libraries, and this type of smartcard technology is regularly used

> **What is RFID?**
> RFID tags consist of an integrated circuit (or chip) for storing and processing information and for modulating and demodulating a signal, together with an antenna.
>
> Passive chips can be read only at close quarters. Active chips contain a power source to boost the signal, increasing their range.
>
> Unlike barcodes they do not need to be carefully lined up in exact positions. This means that they can be read several at a time, anywhere within range of the RFID reader.

for access control in a range of environments. It may control doors or turnstiles by communicating with central controllers to check who is allowed access. It can also record a financial balance or account details, allowing its use in payment cards for minor transactions or in travel cards for public transport systems.

Although RFID chips are becoming increasingly commonplace, the ability of individuals, rather than organizations, to communicate with the chips has been much more limited. This is starting to change, and some mobile phones now have the technology built directly into them. Though this is of a limited scale at the moment, if it takes off, it will allow us to realize a great many more fun and creative innovations. If the RFID tags in our book stock can communicate directly with our users' mobile phones, we will be able to do so much more with the technology.

Probably the key reason why personal RFID readers are not commonplace at the moment is the lack of common standards within the industry. This also makes it difficult for us to create transferable innovations around RFID. It may be best for now, therefore, to work with ideas using RFID technologies that have already been introduced into our libraries; or to use smartcard technology, for which the cards manufactured by MIFARE (http://mifare.net/) seem to be the closest to being a transferable standard.

Using RFID-enabled stock

Due to the proliferation of standards and systems available for RFID-based stock-control systems in libraries, it is difficult to use these systems directly with mobile devices. However, with a little imagination, we could subvert existing systems to do more than the suppliers originally intended.

Using RFID readers connected to cheap computers, such as those that we may already use as catalogue computers in our libraries, we could:

- use our usage data to generate recommendations of other relevant items
- show alternative editions of a book, including e-books, where available
- display user comments, ratings and reviews of items, where available.

These pieces of information could all be texted directly to a phone number, or displayed as a QR codes containing the information, allowing the user to take that information away with them on a mobile device.

These ideas are relatively easy to implement for those libraries that already have RFID-enabled stock, the only additional hardware required being the RFID readers themselves. The stumbling block to implementation is that you need a developer who is able to understand your own library's implementation of RFID and how it communicates to your library's systems. As many RFID implementations are based in academic libraries, perhaps an initial answer would be to pull in student help. If your library has students studying computing subjects, consider investing in a spare RFID reader and offering it to students to experiment with as part of their studies. If they fail to develop anything useful, it will have cost your service only an RFID reader. If they succeed, you may get an additional and innovative service for your library – with the students, of course, getting useful practical experience to use in their assignments, as well as a successful project to add to their *curriculum vitae*.

Using phones as RFID readers

For several years people have experimented by providing users with devices capable of reading RFID tags in order to display extra information or to allow increased interaction with objects. This is especially the case in museums, several of which have supplied visitors with PDAs with RFID that can scan tags on or near exhibits, so as to provide rich additional content.

If RFID readers become more common on user-owned devices such as phones, it could be worthwhile producing mobile apps that scan RFID tags and display extra content, particularly in the sorts of ways suggested above for QR codes. So, for example, we could include the ability to scan library equipment so as to display help materials and videos.

Once RFID readers become generally available on mobile phones, we will also have the opportunity to extend basic RFID services. If we could scan items using RFID applications on phones, we could include RFID functionality in games (scan items for points), allow users to easily generate book lists or recommendations using their mobiles, or build layers of social networking on top of library applications.

Unfortunately, at present there are relatively few mobile phones that include the capability to read RFID tags, but the technology is available and is fairly cheap to embed into current smartphones. Only once mobile phone operators see a key application for which RFID can be used will this functionality become commonplace. Interest in mobile payments is increasing, and it may be that using your mobile phone as an electronic wallet will be the 'killer application' that RFID in mobiles has been looking for. If so, expect it to become the norm on mobile phones fairly quickly, which will give libraries the opportunity to subvert it for their own use.

Summary of RFID

The range of different standards and systems for RFID makes it difficult for libraries to take advantage of the technology at present. With a little imagination, however, they could subvert existing RFID systems in order to add more value for their users. We are also at a point where RFID readers are likely to become increasingly available on mobile phones. If this promise is realized, it will allow the easy embedding of rich content into many objects, as well as the potential to bring Web 2.0 and gaming directly into libraries through specialized applications.

Augmented Reality

Augmented Reality (AR) is the combination of a real, live view of the world around us with a layer of digital information. It is an idea that crops up regularly in science fiction, perhaps with images beamed directly into people's brains, onto their eyes or displayed on contact lenses or glasses.

AR is no longer just science fiction. In order to make AR work, we need a reasonable camera through which to view the world. We need a device that knows where it is and in which direction it is pointing, perhaps using GPS and a compass. We need to be able to connect to the internet wherever we are, in order to obtain information for the virtual part of the view. We need some decent processing power so as to combine information from all of the above, together with a screen on which to display the result. In short, we need what is today a fairly typical smartphone. AR using smartphones may lack the glamour of science-fiction visions of the technology, but there are already many applications available, particularly for Android and Apple phones.

The penetration of mobile phones able to run AR applications (Figure 6.4) is increasing, making this an area ripe for exploitation by libraries with a little imagination. However, there is one significant drawback for libraries – it doesn't work well indoors. GPS is most commonly used for location services, but it relies on clear signals from satellites, which are weakened or unavailable from indoors. Less precise location information is often used in these situations, for example by triangulating positions through using the strength of signals from mobile phone masts. Some projects have used more exact alternatives, such as triangulating positions by the strength of Wi-Fi signals within a building, but this requires detailed mapping and the production of dedicated applications (Figure 6.4).

However, there is a possible way in for libraries, but it is early days yet. Both Microsoft's and Google's mapping services have been working with large indoor spaces such as airports and shopping malls to create maps that can accurately place a user (again using the relative strength of Wi-Fi signals).

Acrossair – www.acrossair.com
One of the first widely available AR apps ('Nearest Tube', showing London Underground stations), the 'acrossair browser' can pull in a range of information and superimpose it on the camera view.
Layar – www.layar.com
The 'Layar reality browser' can pull in a range of information, including many user-generated layers, and superimpose it on the camera view.
Wikitude – www.wikitude.org
The 'Wikitude World Browser' is similar to the two examples above, and again makes it easy to upload information. It started in 2008 on the first Android smartphone, superimposing information from Wikipedia over the camera stream.
ShelvAR – www.users.muohio.edu/brinkmwj/ar/
A library-related pilot project by the Miami University Augmented Reality Research Group! It aims to make it easy to check automatically whether a shelf full of books is in the right order.
Augmented Car Finder – www.augmentedworks.com/carfinder
An example of an incredibly useful specialist application, rather than the more generic AR browsers. I use this app a great deal. It allows you to tag a car with a geo-location; then, when you need to remember where your car is parked, it gives a distance measurement and an arrow pointing towards your car. I never worry about forgetting where my car is parked now, as long as I can remember to use this app!

Figure 6.4 *An illustrative list of Augmented Reality apps*

There is an economic reason for doing this work, because it allows potential users to be located accurately enough for directions to individual shops to be delivered to them, or for offers to be pushed to them, depending on the shops they visit. The potential of targeting advertising, based on a user's behaviour within a shopping mall, provides a tempting possibility of increased advertising revenue for Google or Microsoft. If they can mature the technology in a way that makes it easy for others to do the same, librarians would be able to guide users around their libraries, offering additional information (rather than adverts and offers) depending on a user's position.

The ideas suggested below assume that it will be a while before we can use mapping services reliably to locate a user indoors and are a mixture of some that could be used by any library with minimal time, effort and funding, and others that would require significant resources.

Tag everything you do with your location

Perhaps the simplest idea is to not bother with creating your own AR application, but to make it easy for others to use your information within existing AR applications. Many services that your library may already be using include the ability to tag material with location information. A number of things that you do for your library service using standard Web 2.0 tools could easily include location information for no additional time, effort or money. If you create videos or screencasts, consider posting them on YouTube and including your library's location. If you use Twitter or Facebook for the library service, enable the apps to use location services when sending updates. If you take photos of the library or library activities, include the location and post them on a service such as Flickr. Tag the positions of library buildings on Google Earth.

Many AR applications allow the viewing of information from standard services such as YouTube, Flickr, Twitter or Facebook. So, even without creating your library's own AR layers, your library can automatically appear within many existing applications. Even without your installing a dedicated application or layer for your service, potential users may readily stumble upon your work while viewing Tweets in the area, or looking for pictures near a location.

Tagging your current activities with location data makes it easily reusable by others. It does not depend on your ability to find time to develop services to use this data by yourself, but instead frees up your data to be re-purposed by your potential users using third-party applications in any way they feel works for them.

Use location-aware services

Not strictly AR, services such as Foursquare (https://foursquare.com) are none the less worth mentioning here. Again, libraries can take advantage of these services with little in terms of effort or resources. They allow users of the applications to 'check in' to a location, normally sharing that information with others. For checking in, users receive virtual rewards such as points and badges – and with Foursquare, becoming mayor of a location (for the person who checks in the most during a two-month period). Along with this gaming component, there is also a strong social networking aspect, including the ability to leave tips for other people at a location.

At a basic level, just making sure that your library buildings are on these applications will make them easier for users of these services to discover. You can also leave tips at a location to promote your services. These will often appear in AR applications even for non-users of services such as Foursquare.

In some services, you can register your ownership of a location, allowing you to insert extra information such as opening hours, or making offers available to people who check in. Perhaps waiving fines for the 'mayor' of your location would encourage footfall? Or a free DVD loan for those checked in to your location might achieve the same?

Create your own AR layers

Rather than create your own AR applications, it is possible to do the next best thing and create layers of information that work with existing applications. You can create your own channel with Junaio (www.junaio.com), using tools that it provides. You can create layers within Layar (www.layar.com), using either its own tools or third-party tools such as Positr (www.positr.com) and buildAR (www.buildAR.com), many of which are designed to allow non-technical people to develop these layers. This puts AR within the reach of many librarians who might be keen to use tools such as these but could not hope to dedicate time to acquiring the specialist knowledge to develop them from scratch.

These tools allow you to display any information you wish with a location tag, so you can choose what types of data to include in these AR layers and at what locations they should be displayed. A library service with many branches might display each branch location and basic information. A large library in one location might tag the exact locations of particular collections or services within the building. Whatever you choose to include, it requires time to plan, tag exact locations and promote AR to your users, but no resources beyond staff time are needed to develop it.

Map the interior of your library

This could be done using the standard applications described above, but it is unlikely to be reliable for most buildings, due to the inherent problems of receiving GPS signals within a building. However, it might be worth trying, especially as this is a low cost option.

If it doesn't work, or if you need a more reliable option, applications can be developed using triangulation of wireless internet signals within a building or, simpler still, proximity to low-power transmitters using standards such as Bluetooth. However, either option would require the creation of a dedicated application in order to work. Pilot projects, showing that this is possible, have been introduced within some libraries, such as the SmartLibrary project in the University of Oulu, Finland (Aittola et al., 2003), which combined an interactive map with a location-aware catalogue search to show users the route to a book based on their current location within the library.

With enough money and time, some incredible projects would be possible. Imagine an AR application that shows you where the nearest printers or photocopiers are, that can direct you to books on particular subjects – or, for those of you tired of being asked the question, display the location of items like staplers *before* the user needs to ask us at the enquiry desk.

Combining the idea with other technologies such as RFID tags or Bluetooth-enabled phones, we could even tag our front-line staff, much like the 'marauders map' idea in the Harry Potter books which showed the location of everyone within Harry Potter's magical school. Users could then find the nearest member of staff to help them in the library without having to come to a central desk.

Realistically, however, for most of us, mapping the interior of our libraries is likely to be a task that will only come within reach when the functionality is built into the large mapping services (such as Google Maps) as standard.

A much easier way to map interiors has recently been suggested. While I suspect that this is some way off, Jaewoo Chung at MIT's media lab has developed a system to map a building's interior using variations in the magnetic field created by the structure of the building itself. This should be most accurate in modern steel-framed buildings, but I suspect that those of us who work in libraries full of metal shelves will also be able to map our libraries accurately. It works in conjunction with a badge containing magnetic sensors in order to both map and navigate a building. The demonstration app works with a smartphone containing a projector that projects an arrow onto the floor to show you which way to go. I can't wait until my library is properly mapped and my phone can point me in the correct direction for whatever I want!

Enrich the wider environment with your collections

If you work in an archive or care for special collections, particularly if they are directly relevant to your local area, your resources may be ripe for turning into AR applications. There are several projects around the world that do just this, though they tend to require significant investment in order to be accomplished.

If you can find the time and money to invest in a project that can bring your collections to life in this way, they can be fantastic marketing tools. To be able to walk around a cityscape seeing the skeletons of buildings from plans held in the public library, or to peel back the history of a street to a past age are great applications. To be able to show these using a tiny bit of information held by your library or archive service would be a great marketing tool.

Case study: North Carolina State University Libraries and WolfWalk
Markus Wust and Tito Sierra
NCSU Libraries, USA

The North Carolina State University Libraries has been offering mobile-friendly access to its services and holdings since early 2007, when it launched the first version of a mobile website that included, among other things, catalogue search, opening hours and computer availability; the site was subsequently updated in 2009 to make better use of the enhanced display and processing capabilities of smartphones such as the Apple iPhone and Android-powered devices.

In 2008, library staff began working on another mobile project entitled 'WolfWalk', the goal of which was not to provide 'traditional' library services in a mobile context. Rather, the Libraries wanted to explore how the advanced features of mobile devices, such as location detection, could be used to improve access to and promote the unique and often underused parts of the digital collections.

WolfWalk (www.lib.ncsu.edu/wolfwalk) was initially conceptualized as a location-aware, self-guided historical tour of the North Carolina State University campus. It would allow users to find out about the University's history while being immersed in the actual physical environment they were learning about. While walking across campus, users would be able to view a curated collection of current and historical images of buildings and other significant sites, and read a short history of each site. Access to the information was provided through a location-aware map and a list of all sites on which those in the user's vicinity would be highlighted.

Restricting the project's scope to a relatively small geographical area provided several advantages. It allowed the development team to achieve a high density of content, i.e. to offer users access to large amounts of content in their immediate vicinity, thus improving the user experience. In addition, it greatly facilitated the testing process, which included testing not only for functionality but also for the reliability of wireless network connections and location detection by devices without integrated GPS.

WolfWalk was released in several versions. Initially, development focused on the creation of an iPhone application and a web service that would provide data and images on demand. While there had already been plans also to develop a mobile WolfWalk website in order to make the project available on a larger variety of devices, delays in acquiring an institutional Apple developer's licence emphasized the need for an alternative version. The mobile website version of WolfWalk was released in March 2010, followed by the iPhone application in July 2010. A major update in 2011 saw the release of an expanded mobile website and iPhone application and a new version for Apple's iPad that would optimize the user experience through the device's larger display.

At the same time, the project was reoriented towards a pictorial history of the university. By no longer tying the selection of images to physical sites on campus, the project team was able to include images that were of potential interest to the North Carolina State University community but that were not associated with any particular location on campus. In addition to the already existing map and site list, users could now access photographs based on the decade when they were taken and by choosing between thematic categories such as 'Athletics', 'Student Life', 'Construction', or 'Events'.

Since its original release, WolfWalk has been received with much interest in the library community as well as within the higher education community and has provided inspiration for projects such as the 'BeaverTracks' mobile guide at Oregon State University.[1] During the first 11 months following its release in July 2010, the iPhone application was downloaded over 3,500 times from Apple's App Store. Comparisons of image downloads from the iPhone application and from the website indicate that the iPhone application is used twice as often to view images. The iPhone application's greater use is not surprising, given its availability in and discoverability through the App Store and the expectations of users who (based on comments received after the launch of the website) seem to equate mobile services with App Store applications, to the detriment of device-agnostic mobile websites.

WolfWalk has been a valuable learning experience on many levels, such as technical development, user-interface design and marketing. Among the main lessons for the project team were:

- Make sure that sufficient content or adequate resources are available to create ample content to provide a good user experience. The WolfWalk project was able to draw on, and had permission to use, a large set of already-digitized images. However, it still took considerable time to create a curated sub-collection, to research and author textual content and to build the tools required to successfully manage these materials.
- Build for reusability. Of great value during the development of WolfWalk was the creation of a central web service that could house all of the project's data and deliver them to the project's various implementations, which saved much development time and greatly facilitated data creation and management.
- Invest in infrastructure. In the first version of WolfWalk, adding images involved a fairly labour-intensive process that included the creation of different image sizes and image crops that were then stored on the web server. The 2011 update included a switch to an open-source image server that was already used to deliver parts of the Libraries' digital collections; it could generate all derivatives automatically, as they were requested by the mobile website or iPhone application, and significantly reduced the amount of staff time required for content updates.

Summary of AR

AR applications are accessible by anyone with an up-to-date smartphone and are no longer just science fiction. For a minimal amount of time and effort, we can make our libraries more visible in widely available AR applications, and those libraries with more resources available can create wonderful AR applications of their own.

This area is in its infancy and has much to offer in the near future. However little you plan to do in this space in the near future, it is worth being aware of it. Users are no longer just viewing the virtual world through their mobile phones, or just taking pictures and videos of the real world. They are creating new realities that combine the two. Before long, the ability to take full advantage of AR will become accessible to most of us, rather than to the few.

Linking physical and virtual worlds – summary

Due to the rich variety of electronic content that many libraries hold, for many years now physical libraries have been drifting away from showing the richness of information sources that they hold. With perceived weakening of the connection between the library building and the information holdings, we risk our physical spaces becoming increasingly irrelevant to users. Mobile devices, taking advantage of technologies such as QR codes, RFID and AR, allow libraries to start bridging the gap between their physical and electronic worlds.

There are a range of options to choose from, suiting all budgets and sizes of library service. The case studies and examples provided in this chapter show that some libraries have already started to turn these possibilities into reality. They provide ideas for many of us to start building upon in this exciting new area.

Note

1   http://osulibrary.oregonstate.edu/beavertracks.

Reference

Aittola, M., Ryhanen, T. and Ojala, T. (2003) SmartLibrary – Location-aware Mobile Library Service. In: *Proceedings of the 6th International Conference on Human–Computer Interaction with Mobile Devices and Services*, Glasgow, Scotland, 383–7, www.rotuaari.net/downloads/publication-2.pdf.

Further reading

Chung, J., Donahoe, M., Schmandt, C., Kim, I.-J., Razavai, P. and Wiseman, M. (2011) Indoor Location Sensing Using Geo-magnetism. In: *MobiSys '11, Proceedings of the 9th International Conference on Mobile Systems, Applications, and Services*, ACM, New York, 141–54.
This is the conference paper that presented Jaewoo Chung's work on indoor mapping using magnetic fields. The system demonstrated an accuracy of one metre for 88% of the time – surely enough for us to be

able to navigate within a library, if the system ever becomes mainstream!

Coyle, A. (2011) Interior Library GIS, *Library Hi Tech*, **29** (3), 529–49.
It is still difficult to track location indoors within our libraries. As it becomes easier, Geographic Information Systems (GIS) may become another useful tool in the librarian's arsenal. This article gives an introduction to GIS systems and how they can be used in libraries, including a detailed overview of how they work. A useful GIS primer for librarians who may never have considered these systems before.

Hahn, J. (2011) Location-based Recommendation Services in Library Book Stacks, *Reference Services Review*, **39** (4), 654–74.
This article suggests a model that can be used to create a book-recommendation service for smartphones that uses the user's location.

Kane, D. and Schneidewind, J. (2011) QR Codes as Finding Aides: linking electronic and print library resources, *Public Services Quarterly*, **7** (3–4), 111–24.
The detailed results, including usage figures, of a pilot of QR codes in a US setting. QR codes were put to a range of uses, and the library concluded that it was worth moving from a pilot phase to full introduction into its service.

Lin, H., Lin, C. and Yuan, S. (2009) Using RFID Guiding Systems to Enhance User Experience, *The Electronic Library*, **27** (2), 319–30.
An example of an innovative use of RFID in Taiwan, using the technology to link viewers in a gallery with further information about exhibits. It is easy to see how this could transfer into a library environment.

McCarthy, G. and Wilson, S. (2011) ISBN and QR Barcode Scanning Mobile App for Libraries, *Code4Lib Journal*, 13, http://journal.code4lib.org/articles/5014.
A slightly different take on QR codes and scanners, this article describes how a university embedded a scanner into its library app. The scanner could be used on any book to search for holdings by ISBN. It can also be used to scan QR codes used within the library and its services. This approach is how QR codes are most likely to be successful – by embedding scanners into apps that are useful for a range of activities, rather than expecting users to know what a QR code is and select a suitable scanner.

Ramdsen, A. (2010) *The Level of Student Engagement with QR Codes: findings from a cross institutional study*. Working paper. Bath: University of Bath, http://opus.bath.ac.uk/19974/.
This working paper shows the level of student awareness and engagement with one particular technology in a UK university that has

experimented a great deal with QR codes. It shows the level of student engagement growing from an initially very low base and includes a large number of qualitative statements from survey respondents. A really useful place to start if you are considering introducing QR codes into your own institution.

Walsh, A. (2011) Blurring the Boundaries between Our Physical and Electronic Libraries: location-aware technologies, QR codes and RFID tags, *The Electronic Library*, **29** (4), 429–37.

A confession here: this is one of my journal articles. It is directly relevant though! I go into more detail on the potential for the technologies discussed in this chapter to blur the boundaries between our physical and electronic libraries. As with all my journal articles, my final version (i.e. before typesetting) is hosted on my university's repository and is easily discoverable using any major search engine.

7

Mobiles in teaching

Introduction

There still seems to be a culture of 'turn off your mobiles' within most educational establishments. Most of us work in places where it is the norm for the people whom we teach to possess a mobile phone, often a smartphone – a state of affairs that will only increase in future.

Instead of taking advantage of these wonderful, portable computers, we deliberately try to prevent others from using them. Rather than using them in teaching, we tell our classes to turn them off, ignore them, try to stop them interfering in 'our' teaching. Instead of regarding them as interfering with your teaching, why not take advantage of them? Why not draw them into the process and let them enhance our students' learning?

Should we use students' own devices, or provide class devices?

There are two approaches that we can take here. We can buy sets of devices for class use, similar to those that may be owned by members of a typical class anyway, or we can use our students' own devices. Both approaches have benefits and drawbacks.

The key advantage of using students' own devices (normally phones) is that the cost of providing and updating the equipment doesn't fall on the library and we don't need to worry about how much a set of devices for a class might cost. If most of the members of a class bring mobile phones with them, particularly if these are smartphones, the equipment will cost us nothing. There is the added advantage that the students will be familiar with the devices – after all, they are their own phones! You may need to spend time explaining how you intend to use them, but no instruction will be needed in the operation of the hardware, the actual devices.

However, we must be aware of several disadvantages before we start using the students' phones. Firstly, not everyone may have a phone or be willing to use it in class. This is especially true if there are concerns about costs (e.g. the cost of sending text messages) or about privacy (e.g. about the retention of phone numbers). You must be up front in dealing with these issues if you want to use the students' own devices in class. If you want them to send or receive text messages, make it clear from the start how much this is likely to cost (the costs will generally be acceptable once the students know). Make sure that you don't keep mobile phone numbers unnecessarily and tell members of the class that this is the case. Whenever you do an activity using students' own mobile devices, unless you are sure that everyone is happy to use their own phone, try to ensure that people can respond as a small group. This way, you need only one person in each group of two or three to use their own phone.

The second issue concerns using a range of different devices in the classroom. With mobile phones you can be sure that the devices can all send text messages and make phone calls, but any additional functionality (such as taking photos, videos, access to the internet etc.) can vary greatly. Further, you can't rely completely on the most basic functionality. I once checked the mobile phone signal of two different mobile networks in a lecture theatre. Happily, both were working. I tried to use text messaging during the class. Unfortunately, a large proportion of the class used a different mobile network that I hadn't checked and that didn't have a usable signal in the room, thus excluding many members of the class from the planned activity. So, using students' own mobile phones will always be inherently more risky than using a set of identical devices that you are confident will work; but cost (to the library) and familiarity (to the students) are powerful advantages that will overcome these problems for many of us.

The key advantage of buying and providing class sets of equipment is that you can be so much more confident that they will work. We can test and practise with them beforehand. We know how many devices we will have for use in the class and exactly what functionality they have. However, classrooms, libraries and IT store rooms around the world have obsolete sets of equipment purchased for a project, used for a short time, then stored and forgotten about. If we are lucky, we may get funding to buy sets of mobile phones, handheld games devices, video cameras and who knows what else for a project. Most of us are not that lucky, but even for those who do get to buy exciting sets of equipment, updating and replacing them is a different matter. Buying and maintaining up-to-date handheld devices to use in class is simply not an option for most libraries because it is too expensive, and hard to justify.

Buying class sets of equipment will also mean that we are likely to have to provide instructions and guides for using the equipment, and this will eat into valuable teaching time. Here, some valuable lessons to learn are to make sure that if we go down this route, we do so in the best possible way. In the UK there have been 104 projects supported by the MoLeNET initiative (see www.m-learning.org/case-studies/molenet- and websites of individual institutions), and these are a rich resource of case studies and practical experience. Take advantage of the experience gained by others, such as the MoLeNET projects. These projects have had the opportunity and the funding to experiment. Learn from the inevitable mistakes and great successes of others, rather than repeating the same mistakes yourself.

So there are pros and cons both to using learners' own mobile devices (particularly phones) and to providing class sets of equipment. This chapter will give examples of both options for a range of possible activities.

Case study: Mobile technology and information literacy instruction
Robin Canuel
Humanities and Social Sciences Library, McGill University, Canada
Chad Crichton
University of Toronto Scarborough Library, University of Toronto, Canada

In 2011, the teaching work of academic librarians is largely focused on making sure that our students become technologically savvy and information-literate lifelong learners. When librarians talk about information literacy they are usually simply referring to the set of skills needed to find, retrieve, analyse and use information. However, it is worth noting that the ubiquity of mobile internet access and the unique technological capabilities of today's mobile devices have, arguably, broadened the requisite skill set that an individual would need to demonstrate in order to be considered information literate in a modern context.

With a mobile device in hand, a researcher not only has access to an internet connection, but also has a variety of hardware at their command that they can use to engage in their research: a microphone that can be used to input terms into a database using one's voice, or to search for metadata about a song by letting a search engine listen to the tune; a camera, which can allow visual inputs to be used for searching (as with the Google Goggles app, or for accessing information by scanning barcodes or QR codes); and a GPS chip, enabling searching on the phone to be 'location aware' and search results to be ranked by their distance from the searcher.

Many applications make use of more than one of these unique capabilities, such as modern Augmented Reality apps, which combine a mobile device's GPS location information and the viewfinder of the device's camera to give a searcher a 'heads-up' display highlighting available information about their surroundings and literally augmenting the reality surrounding the user by superimposing digital information on top of the real world around them. Mobile technology has changed *how* people search for information, and it is essential that academic librarians embrace the challenge of educating their students about the changes in the methods employed to access information and their implications for research and lifelong learning.

The rapid spread of mobile technology is as much an opportunity as it is a challenge for those of us in the field of information-literacy instruction. Students, particularly younger undergraduate students, already use their mobile technology frequently in their daily lives, and so it affords us the opportunity of reaching out to these young researchers in a context that they are already familiar with, to some extent, and that is potentially very engaging.

The new ways of searching also provide librarians with new avenues for explaining traditional information-literacy concepts pertaining to the acquisition, examination and use of information. For example, a discussion with students about how to choose the best apps to fulfil their information needs provides an opportunity to discuss the importance of metadata (taking advantage of app store user comments and rankings) and the importance of turning to the appropriate source to address one's research needs.

Similarly, discussing with students why the Google Goggles app is capable of returning search results based on a captured image of the Eifel Tower, while it offers no results when they take a picture of their local public library, emphasises the relationship between search inputs and result outputs, a relationship that is not always obvious to students so used to Google's tendency to always bring them *something* in response to *any* search query (whether that 'something' is appropriate to their needs or not).

One of the first steps in incorporating mobile technology into one's information literacy endeavours is to build awareness of the availability of mobile resources and sites through the creation of a library mobile web presence, be that a site, an app or both (e.g. the sites for the McGill Library and the University of Toronto (UofT) Libraries[2] and the UofT Libraries mobile app http://m.library.utoronto.ca/mobile.html[3]). In response to the proliferation of mobile-friendly library resources from our vendors, instruction librarians must also promote and use (or create, where the need exists) subject guides focused on mobile resources available to researchers in their field.[4] With the explosion of mobile apps, the 'collecting' and curating of mobile apps for researchers is also a highly valuable service.[5]

The primary challenge of incorporating mobile technology into information literacy instruction is, of course, the availability of the hardware itself. Some students do not have a personal mobile device. Even when one is lucky enough to have a class of students who all own a smartphone, there are still differences between the various devices in terms of hardware, functionality and the availability of applications. One possible solution to the problem of device specificity is simply to provide institutional devices to the students. This is an area currently being explored by a number of institutions, particularly in view of the emergence of tablet computing.

Expanding on the practice of loaning e-readers to patrons, many academic libraries now offer iPads for loan to their faculty and students. For teaching, some institutions have begun to purchase, or are planning to purchase, multiple tablets or other mobile devices for classroom use by their students. Ironically, the increasing popularity and availability of e-readers and tablets, not necessarily the first hardware that one thinks of when discussing 'mobile technology', has added urgency to the push to incorporate mobile technology into our instructional efforts. Today, libraries that might never have considered buying a lab's worth of iPod Touches for teaching mobile information literacy are none the less making plans to purchase iPads for the classroom.

Academic libraries are now developing and providing workshops to help students to enhance their mobile information-literacy competencies. For example, librarians at McGill University currently offer a workshop entitled 'McGill Library from the palm of your hand', to teach students how to access and use a variety of library resources available for smartphones, e-readers and tablets. Academic librarians everywhere are discussing the functionality of various apps for teaching and are exploring creative ways to make use of apps created by external developers both inside and outside the classroom. With such a rapidly developing technology it is important to teach students to take advantage of all the specialized search options specific to mobile devices and to navigate the ever-expanding collection of mobile resources and applications.

Using text messages to teach
Polling

There is a long tradition, particularly in the USA, of using handheld devices to poll members of the class. These are normally dedicated interactive handsets or 'clickers'. They can make it easy to create a decent level of interaction with a class, particularly if the activity is well planned and properly integrated. They can help you to check knowledge, facilitate discussion and even formally test members of a class.

There have been moves towards installing software on other mobile devices in order to replicate the benefits of interactive handsets. Some institutions have even supplied all students with mobile devices that allow them to integrate this function throughout their teaching.

There are commercial applications that can be installed on smartphones and devices such as iPod Touches in order to turn them into the equivalent of these traditional interactive polling handsets, including some applications produced by manufacturers of 'normal' clickers. However, there have also been moves to produce free versions, including a European Union-funded project, which may be worth investigating.[1]

A third way of using mobile devices for polling in our teaching is to use text messaging (SMS) from students' own mobile phones and, for those

PollEverywhere –
www.polleverywhere.com
This is an audience-response system with a range of options that can work via text message (with a range of international numbers), Twitter or web voting. It makes it easy to create polls embedded in PowerPoint. Example slides and instructions for its use in your own presentations are provided. It is free for small classes (currently 30 voters per poll), with paid-for versions for increased functionality and larger class sizes.

Votapedia – www.urvoting.com
A system developed for free educational use in Australia, this has a range of functionality and allows voting by text message or web browser. The code is open source and is available for others to repurpose.

SMS Poll – www.smspoll.net
This is an audience-response system with a range of options that can work via text message (in Australia, New Zealand, the United Kingdom and the USA) or web voting. It makes it easy to create polls and embed them in PowerPoint. It is free for small numbers of polls and limited responses, with paid-for versions offering increased functionality, larger class sizes and more polls per month.

TwtPoll – http://twtpoll.com
A range of question types can be asking using this polling software, which is free for casual users. Paid-for versions are available with increased functionality and for larger numbers of respondents. It is designed to work with Twitter and allows simple and easy web voting.

Figure 7.1
Some alternatives to interactive handsets ('clickers'), using online and mobile technologies

libraries that are willing to use users' own mobile phones in teaching, this can be a valuable alternative.

There are many services now available that can use text messages combined with online collation of votes (Figure 7.1). These systems can often replicate the functionality of the traditional interactive handsets or clickers, but at a fraction of the cost. They are often free to use for small numbers of students, with a subscription being required to bring in extra functionality or to allow more respondents to a poll. Even with a subscription, because no hardware is required, these systems can be considerably more affordable than dedicated clickers.

More uses for text messaging

Text messaging is limited in many ways. Messages are short (160 characters), restricting the amount of information that can be sent. There can also be cost implications, which, though small, can restrict the use of text messaging.

Considered as a supplement to other teaching methods, however, text messaging can be a useful tool that is available to nearly everyone. During classes text messages can be used in a similar way to Twitter, for instance, as a back channel, especially if you pull messages together using a text wall.

However, after a class has finished may be when text messages provide the most added value. The short length of a text message makes them ideal for sending out short reminders of key facts, reinforcing knowledge retention.

Text messages could be used after library inductions, reminding users of

Text walls and similar
Many organizations offer text walls, which collate and display text messages on a website. A UK educational example is www.xlearn.co.uk/sms.htm, which currently costs just £25 a year for teachers.
Microblogging services such as Twitter (www.twitter.com) and Moblog (http://moblog.net) normally include the ability to post to a wall via text message and are designed to cope well with short messages, so can easily be used as text walls.

key facts (such as how to renew books) at an appropriate time. They could remind learners of key resources covered in your teaching sessions. They could even just be used for simple but valuable tasks such as reminding people where they can obtain additional help after a session has finished.

Recording activities (video, audio and pictures)

Many handheld devices, from games consoles to music players and, of course, mobile phones, typically include a camera and some way of viewing images. This functionality often includes the ability to record and view videos.

No longer is it difficult or expensive to record learning activities as they happen. It is now quick, cheap and simple to take pictures or record videos during teaching and make them available online.

In the past, the feedback to the whole class from a teaching activity or discussion would have been oral (and lost instantly) or written (and largely inaccessible after the class). Mobile devices now give us alternatives.

The results of discussions can be recorded and put online instantly, allowing easy access after the event. Whether using open hosting environments like YouTube or Flickr or closed ones like an institutional Virtual Learning Environment, discussion can continue after the formal class has ended (Figure 7.2).

... and now for the news
This is a sample learning activity using any device with internet connectivity and a video-capable camera, just one per group within your class.

- Set up an account with a video-sharing site such as YouTube.
- When students have been working together in groups, at the end of an activity ask them to structure their results, thoughts or findings as a 60-second news report.
- Make sure that each group has access to a mobile device with a video camera and explain that they have 10 minutes to write a news report, rehearse it and record it.
- Explain that news normally goes out live, so they will not have a chance to watch the video in their groups and then edit or improve it.
- Make sure that they are all starting to record two minutes before the time is up.
- Collect the video from each group, preferably by the groups uploading their results directly to a video-sharing site using the account that you previously set up.
- Tell them how to find the videos and encourage them to watch and comment on them after the class has finished.

Figure 7.2 *An example of a learning activity using the video capability of a mobile device*

Closely linked to video recording is the ability to use mobile devices to create instant audio recordings. This is particularly the case for mobile phones, which, after all, are essentially a portable microphone combined with a means of transmitting the sound.

Many free tools are available to create audio recordings and podcasts. Free audio-editing software can be used to tweak recordings if necessary; or just record audio live and broadcast it automatically using tools such as those listed in Figure 7.3.

AudioBoo – http://audioboo.fm
This is freely available on various mobile platforms, over the web and, originally, by normal landline phone (by ringing a set number, though this functionality may not be available at present). You record a message, add some simple details such as the title of your 'Boo' (recording) and allow AudioBoo to publish it. You can easily provide people with an RSS feed of your recordings.

iPadio – www.ipadio.com
Very similar to AudioBoo, this is another free tool that makes it incredibly easy to create live podcasts. It currently seems to offer more editing options than AudioBoo and longer podcasts.

Audacity – http://audacity.sourceforge.net
This free, easy-to-use software is available for a range of operating systems and can even be run from a memory stick if you can't install the software on computers at work. If you record audio on a device and want to edit it before publishing, this is an excellent free piece of software to start with.

Figure 7.3 *Some free tools for creating audio recordings and podcasts*

If taking pictures, recording audio and video are now relatively easy, it is even easier for people to consume such material on mobile devices.

When creating your standard online teaching materials, remember that people may wish to consult these materials on the move. Consider making them available in mobile-friendly formats and try to make sure that they are usable on devices with small screens (Figure 7.4). When first creating videos suitable for mobile devices it can be hard to navigate the range of different formats, as there is a confusing mixture of standards, none of which seem to be viewable on all devices. If your library is willing to host materials on third-party websites rather than its own pages, services such as YouTube are sophisticated enough to ease many of these problems for you. YouTube will try to detect the type of device being used to view videos and deliver them in an appropriate format, doing much of the hard work of worrying about formats for you.

- Remember the size of the typical mobile screen!
- Zoom in on detail that you wish to show.
- Any text that you add should be short and in a large font.
- Make a transcript available.
- Formats can be difficult.
- Host videos on a site than can automatically deliver them in different formats.
- Flash does not work on Apple devices! It is also unlikely to be supported on Android devices in future, with HTML 5 being the preferred mobile standard.
- Size matters.
- Keep the file size down in order to speed download to mobiles.
- People are unlikely to watch long videos on a mobile device. Try to keep them to between one and two minutes long.

Figure 7.4 *Some tips for creating mobile-friendly video recordings*

Shared activities

Mobile devices, especially personal ones, are great for sharing. Even for those of us who are reluctant networkers, mobile devices let us share easily and painlessly with our colleagues and fellow learners. Take advantage of this easy sharing in your teaching to get conversations and some high-value group learning going.

At my young daughter's dance class several of the children have Nintendo DS devices. You see them sitting, seemingly alone, in different corners of the room, absorbed in their handheld games devices. When you peer over someone's shoulder, however, you find that they are far from alone. They are writing messages to each other, or playing joint games where all their consoles are linked together. Why not use this in the classroom? You can easily create quizzes that would run on these devices, or let groups take part in a full-class discussion, writing down their thoughts on a shared document using a handheld games machine.

Perhaps the easiest cross-platform way of encouraging discussion via mobile device is through an existing social network, such as Twitter. Twitter is easy to update from any device that can connect to the internet, as well as by text message. Messages can be pulled together using 'hashtags' (a label prefixed by the symbol # and often agreed on beforehand), allowing discussions from many people to be displayed in one place. Twitter can be used as a back channel for feedback, letting learners talk amongst each other and allowing you to listen in and gain extra information about the learners' needs and your

> **What is Twitter?**
> Twitter (www.twitter.com) is a free social network or microblogging service that allows people to post updates of up to 140 characters in length.
> You can choose people to 'follow', which means that their updates will appear in your timeline, as well as search for updates posted by anyone, using user-generated tags (hashtags).
> The short length of updates is to allow updating by text message, though most updates now tend to be made using dedicated apps on smartphones.

Twitter treasure hunt

This learning activity uses Twitter to run an interactive treasure hunt.

- Design a treasure hunt-style activity. In libraries this may be to discover elements of the library service for an induction activity.
- Split a class into groups, making sure that one member of each group has a Twitter account (it takes only a few seconds to set one up) and a mobile device from which to send updates. Assign a hashtag to the activity so as to make it easy to collate the updates into one place.
- 'Follow' each Twitter account that will be taking part in the trail and ask the groups to follow your account.
- Give out a clue that leads to the first point of the library trail or treasure hunt. Ask the groups to tweet the answer, plus hashtag, when they find it. If you want to conceal the answers so that other groups can't eavesdrop, ask them to Direct Message the answer to you.
- Monitor Twitter, and when a group finds the answer, tweet back (or direct message) the next clue.
- You can tweet additional clues if it appears that people are struggling.

Figure 7.5
An example of a learning activity using Twitter

performance. It can be used to pull discussions together into one place online, especially if you have a geographically dispersed class. You can even use third-party tools that integrate with Twitter so as to allow functionality such as polling. With a little imagination, it can be used to bring fun and games into classes – perhaps enlivening them by using treasure hunts (Figure 7.5).

Library trails

A common induction or orientation activity is a library trail. Mobile devices can really come into their own for these activities. Besides the example of Twitter (Figure 7.5), you could use text messaging to release one question at a time, or bring in location awareness to prevent cheating.

Scavenger hunts are a popular activity on GPS-enabled mobiles and have grown in popularity since smartphones have become common, and so many applications are available. Several of the applications include the ability to take and upload pictures, or to scan QR codes to prove that users have found the correct location or answer to a question.

This idea can easily be adapted for library trails, which must surely be more fun – and less time consuming – than leading tours yourself!

Case study: Location-based gaming: meeting the overwhelming demand for school visits

Linda Barron
Client Services and Collections, State Library of Queensland, Brisbane
Queensland, Australia

Introduction

Over the past two years the State Library of Queensland has seen an increase in the demand for visits to the Library by school groups. With a small group of staff dedicated to learning, it became apparent that tours guided by library staff were no longer practical. Staff decided to investigate platforms, software and devices for the development and delivery of a self-guided tour for schools. These tours need to be engaging for the students as well as educational. Groups are eager to learn what the Library has to offer in terms of expertise in study and research, as well as about the resources that it contains. Several location-based gaming options were investigated, one was selected and a self-guided tour was developed, tested and implemented. This case study shares recommendations and learning from this investigation and implementation.

Location-based gaming

A location-based game is one in which the game play somehow evolves and progresses via a player's location. Thus, location-based games almost always support some kind of localization technology such as a global positioning system (GPS) (Wikipedia, 2011). Scavenger apps, for the most part, make use of satellite positioning like GPS.

Scavenger apps

Do you remember doing scavenger hunts around the streets of your home town? You were given a set of clues and went from place to place performing tasks such as finding someone, counting steps, noting down the date when a building opened and finding other locations around the town until all the clues had been solved and you reached the end. As mobile phone technology has evolved, apps have been developed that enable scavenger hunts to be completed on smartphones.

There are a number of scavenger apps available in the iOS and Android market. The State Library looked at three apps and their versions. These were:

- Scavenger Hunt Classic
- SCVNGR
- SCVNGR Museums
- Scavenger Hunt Lite
- Scavenger Hunt Deluxe.

The app that we selected on which to develop our library tour was SCVNGR. This app was selected because it best met the criteria that we had set for the development of a self-guided tour on an externally developed app.

Scoping the project

The Learning and Participation and Literacy and Young People's teams had kept detailed records of previous visits by school groups to the State Library. Along with this there were also records of groups that the Library had been unable to accommodate. This assisted in identifying the target market. After analysis, it was ascertained that there were three specific groups interested in self-directed tours:

- years 5–7
- senior students – years 10–12
- family groups.

The most requests for Library tours came from year 5 through 7 school groups. This group was particularly interested in finding out what the Library is all about.

School groups from years 10 through 12 are where most requests about study and research skills come from. These senior-class groups are interested in finding out how to use the Library, what we have here and where everything is.

Families were also identified as a potential group to make use of a self-guided tour. Many of the school holiday activities that involve making your way around the Library, following clues and finding resources, have proved to be extremely popular and they encourage this user group to explore library spaces.

Accessibility was another consideration when selecting the app. Clients should be able to access the app on an Apple or Android product. The app should be free to download. As the State Library has free Wi-Fi internet, clients would not need to use their own data plan to download the app. Printed copies of the activities would be required for those who did not have a device that was GPS enabled.

As GPS locating is sometimes not 100% accurate, it was important that the app did not need exact locations. GPS can be problematic, because on occasion the State Library is not at the State Library.

Other considerations included the app's needing to have the capability to cope with a number of people participating in a hunt at one time. At this stage of the project there were no funds set aside for an on-going subscription to a platform. As such, whatever was selected needed to be free. As the State Library was looking for something to replace staff-directed tours, the product selected would need to be sufficiently intuitive that clients could use it with ease.

After researching and testing a number of apps, it was decided that SCVNGR was best placed to meet our requirements.

SCVNGR

This is possibly the best-known scavenger app. You can download the app to your GPS-enabled device and participate in location-based hunts that have been developed by others. Hunts developed using SCVNGR can include images, and participants can upload images as clue answers. The app enables QR codes to be scanned, allowing additional information to be pushed out to participants. Prizes and rewards can be added when you are developing hunts, which is something worth considering if you have the resources. Participants can choose to share their participation in a hunt through a number of social networking sites.

The app is free to download, develop treks and challenges and participate in hunts. SCVNGR is available to use on both Android and Apple products. The treks and challenges do not need to have an absolute location. Along with this, there is no limit to the number of participants that can use a hunt developed on SCVNGR.

In order to design hunts you need to sign up to the SCVNGR website as a 'builder'. Building hunts is best done on a computer or laptop, as the app does not allow building within it. Builders can design as many treks and challenges as they wish; however, the free option allows only five to be active at any one time. Analytical reports can be generated and exported from the desktop version. These show useful user statistics about treks and challenges.

The development stage

Fifty per cent of the Australian population own a mobile phone that is internet capable (Timson, 2010). To ensure that those without smartphones can participate in a self-directed library tour, printed copies of the tour have been made available through the State Library's website. It is worthwhile noting that during the trial of the self-guided tour on the app the students had sufficient phones to have one between two students. Based on anecdotal reports, it is expected that this trend will continue.

The first step undertaken in the development stage was the compilation of written scripts. Of greatest urgency was putting together self-guided tours for school groups. Years 5–7 were extended to cover years 5–9, so as to align with the Queensland Education Department's middle phase of learning. This decision was also influenced by the fact that there was a gap in activity for these year groups.

Once the scripts had been developed, the activities that were to be part of the scavenger hunt were extracted. Whilst developing the *trek* a number of things were learnt that it is useful to share:

- *Challenge* is the default tab in SCVNGR builder; however, if you want to develop a hunt you first need to set up a *trek* and then add challenges to the trek. Challenges cannot be attached to a trek via the challenge tab.
- As the tour of the State Library was to explore the floors, the *title* of each challenge became the floor level, making it clear where participants should be located.
- The *challenge, answer* and *message* blocks allow a maximum of only 140 characters to be entered.

- The free version gives you a capacity to make only five activities live. As a *trek* is required in order to add *challenges*, you can have only four *challenges* within a *trek*. There is an option to upgrade and pay a monthly fee.
- There is no limit to the number of *treks* and *challenges* that you can build. If you choose the free version, you can activate only five of these.
- Each *challenge* within a *trek* needs to be made active.

The trek and associated challenges were made active and the next phase was to undertake testing so as to check usability and content.

Implementation

Testing was undertaken first by staff and then by a large group of students. The student group consisted of 75 year 9 students. As previously mentioned, the group had a sufficient number of mobile devices to have one between two students. Devices were a mix of iOS and Android, allowing testing to occur on a number of different devices.

Observations from the staff test and student group revealed that:

- The challenges are listed under place rather than title. If conducting a trek over a number of locations, this would not be an issue. It was observed that students undertook challenges in the order in which they were designed.
- The images added to each trek assist participants to find the location where they need to be.
- Once participants have pressed *Go* within the *challenge* they cannot go back to the answer box to enter further information. This would not be a problem if only one question was asked in each *challenge*. The State Library *challenges* have multiple questions, taking maximum advantage of the five free activities.
- As the answer to the questions within each *challenge* was designed to be an open response, students were unsure whether they had got the answer correct. Most questions were easy to respond to and students had little difficulty answering them. The group leader (e.g. teacher) needs to read the information sheets provided to support the trek and to ensure that pre and post activities are undertaken. Those who find the trek themselves, without the official introduction, may not have the same advantage.
- If a location or clue was hard to find, this was overcome by participants' seeking out staff to ask for assistance.

After the testing phase all information pertaining to the self-guided tours was made available on the State Library's website. It can be found at the following location: www.slq.qld.gov.au/info/teach.

It should be noted that when open responses to *challenges* are allowed, participants can answer however they choose. This feature is not quick to remove because the builder is unable to do it themselves. SCVNGR support must be contacted and informed of the exact place where the comment was left and will then remove it. This can take several days.

Conclusion

The State Library has now launched the education-focused self-guided tour, available as an app or as a printable PDF hand-out. The app version, suitable for years 5–12, shows floor plans of the State Library, giving an idea of what can be found on each level. The SCNVGR app allows school groups and other interested members of the public to take a tour of the Library through a series of challenges. The app has enabled the development of a tour that is both engaging and educational and that is appealing to the target market. The self-guided tour has increased the State Library's capacity to deal with the level of demand for school group visits.

References

Timson, L. (2010) *Australians Take to Mobile Internet*, www.smh.com.au/digital-life/mobiles/australians-take-to-mobile-internet-20100429-tszn.html.
Wikipedia (2011) *Location-based Game*, http://en.wikipedia.org/wiki/Location-based_game.

A window to another world

In environments where Wi-Fi is available and reliable, such as many of our libraries and teaching environments, free video calling becomes a realistic option. Services such as Skype (available for most platforms), Facetime (on iOS devices) or Google Plus hangouts (most platforms) can use a mobile device's camera, microphone and Wi-Fi connection to set up free video calls to other devices.

These could be used as a way of including geographically dispersed members of a class in a teaching session, enabling a physical class or seminar to be delivered simultaneously to distant attendees. While not ideal as a platform for delivering distance-learning classes, this can bring in extra people who might not otherwise be able to attend.

Perhaps a more interesting way of using these services is as a window into another world. As video calling is free on these platforms, it can be left running as long as you like. Try experimenting in your teaching by fixing a tablet computer to a wall, linking it to another device via Skype or Facetime and treating it like a 'magic window' into another world.

You could give the linked device to a colleague in another location and set up scenarios within your classes in which your colleague would be called upon (especially if they were willing to indulge in a little role playing). You could set exercises in class that involved interviewing a subject expert; create quizzes and learning games in which some of the clues are available by asking the 'magic mirror' (mirror, mirror on the wall ...); or even a have quest to free the person trapped within the 'magic window' on the wall (once all the questions have been answered, the person is freed to come physically into the room). Using these video calling services on mobile devices, together with the imagination to see the devices as windows onto another reality, provides a whole new set of ideas to use in teaching.

Alternatively, use paired devices with groups carrying out treasure hunt-style quizzes. One half of a group can carry out 'desk research' to solve the problems or clues, and the other half can follow their instructions in order to discover whether they are correct and reveal the next clue. With a mobile 'window' open between them, the two halves of the group can readily communicate with each other, creating something that it would be difficult to imagine without these mobile services and devices.

Summary

Mobile devices, whether the students' own or supplied by your library, can be a valuable addition to the teacher-librarian's toolkit. Whether using text messaging, these devices' cameras and/or audio-recording capabilities or just internet connectivity and processing power, the uses to which you can put them are numerous.

Remember, however, that they are just a tool. A fun, flexible, valuable tool, but a mindless tool none the less. As such, you must plan their use properly and use them well: they are not a substitute for good pedagogy. Planned well into your teaching activities, they can add real value and bring a new dimension into your teaching.

The final case study in this chapter shows how creative you can be in using mobile devices in teaching. The use of mobiles is integral to the exercise, but great creativity, planning and design of the activity made it a success.

Case study: Adventure night: interactive storytelling at Box Hill School
Sarah Pavey
Senior Librarian, Box Hill School, UK
The amount of teaching and learning using mobile devices in the classroom has grown over the last year at Box Hill School and we are constantly seeking ways to enhance

their use as an educational tool.

It was decided to base our summer interactive story evening on an Indiana Jones adventure theme. We wanted to include the use of mobile phone technology, and it seemed an ideal opportunity to incorporate Quick Response (QR) codes into the plot.

The event began with 15 new recruits (boys and girls) being shepherded into the briefing room at the Library Operative Special Team (LOST) headquarters (an organization so secret that no one had ever heard of it). The recruits, dressed in adventurers' costumes, heard that the Auto Reactive Kaos device (ARK) had been stolen and hidden somewhere on the campus and that their mission was to recover it safely. Clues would be given through secret codes and they would be guided by the two Mrs Librarys. However, Russian soldiers were also keen to retrieve the ARK, and so a race was on!

The agents were each given an ID card on which was printed a QR code. When this was decoded using their handheld transmitter (aka mobile phone), their code name was revealed.

Soon it transpired that the evil Mr Mink was responsible for the loss of the ARK, but in his hasty escape he had dropped a QR code clue at LOST HQ. When this was found the adventurers were directed to the high-ropes agility course.

The second clue proved more difficult to decipher, and so the Mrs Librarys took it back to HQ for some research – a neat excuse to retrieve the fish-and-chip supper! The agents then watched *Indiana Jones and the Temple of Doom* and kept an ear and an eye open for tips that they might learn from the plotline.

By now darkness had fallen and the second clue pointed the agents in the direction of the tennis courts, where, unfortunately, they were ambushed by the waiting Russians. Our hapless agents were blindfolded and kidnapped in the waiting bus. It sped away down the dark lane.

But a counterattack was mounted by the ace team from LOST and the agents were freed one mile from the base. They jogged their way back under the stars.

At LOST another QR code had been found. It led the agents to an old man in a dark, unlit shop who possessed a gold medallion. He would sell only if the agents delivered him gold coins.

The old man handed our agents their fourth QR code clue.

It led them to the tomb of a medieval crusader knight who stood guard over the gold coins. Three by three, in turn the agents climbed the staircase in the pitch dark. The knight rose from his coffin and anointed them with water from a relic bottle of the Virgin Mary (something he'd picked up in Lourdes!). Then the agents had to plunge their faces into a bowl of special powder (flour) in order to retrieve the coins with their teeth.

The knight warned the agents that when they left they must go straight back to LOST HQ to retrieve the next clue, and they should not speak to the other agents waiting outside. The agents fled down the stairs, screaming.

Back at HQ the agents were given a tutorial on disarming bombs. This involved placing a metal strip across two electrodes to create a circuit and stop the clock ticking. The fifth clue was hidden inside an old, leather-bound book. It guided the agents to the drama room, where dramatic events would certainly overtake them.

Sure enough, while the agents were exploring the drama room there was a burst of fire and the Russians emerged with a ticking bomb. Luckily, the LOST agents were well prepared and quickly made it safe. Pasted to the bomb was a final QR code clue.

This directed the agents back to the knight's tomb. After climbing the dark stairway once again they saw the ARK shining, but surrounded by dead Russian soldiers. The agents took the ARK back to where it belonged – a satisfactory end to the tale as the clock struck midnight.

The participants in the event were then asked to produce an account of the evening as a creative writing exercise.

Overall, this was a fun way to engage our students with the applications available on their mobile devices. They were able to see how QR codes can be used within a learning environment to link to simple messages, websites and pictures. With the use of QR codes becoming more familiar in the community as a means of communication, we wanted our students to gain an understanding of how this technology works and how it can be applied in practice.

Notes

1 http://scom.hud.ac.uk/scomzl/joan-research/SRS-pub.htm.
2 http://m.library.mcgill.ca and http://m.library.utoronto.ca.
3 http://m.library.utoronto.ca/mobile.html.
4 For example, http://m.library.mcgill.ca/columnview/healthsciguide.
5 For example, http://guides.library.utoronto.ca/e-reading.

Further reading

Berk, J., Olsen, S., Atkinson, J. and Comerford, J. (2007) Innovation in a Podshell: bringing information literacy into the world of podcasting, *The Electronic Library*, **25** (4), 409–19.
A detailed case study from an Australian university that introduced a range of podcasts in order to help it to improve students' information literacy. A simple mobile technology that is cheap and easy for most of us to both produce and use.
Habel, C. (2011) VotApedia for Student Engagement in Academic Integrity Education, *ergo*, **2** (1), 15–25.
An example of how mobile phones can be used as voting systems, in this case using VotApedia in Australia to teach subjects such as plagiarism.
Havelka, S. and Verbovetskaya, A. (2012) Mobile Information Literacy. Let's use an app for that! *College & Research Libraries News*, **73** (1), 22–3.

A fairly short piece that outlines how the authors are starting to consider mobile information literacy and how they should be teaching these skills to their library users. The paper is short but interesting and covers an issue that perhaps more of us should reflect upon when considering information literacy and mobile devices.

Kazakoff-Lane, C. (2010) Anything, Anywhere, Anytime: the promise of the ANimated Tutorial Sharing Project for online and mobile information literacy, *Journal of Library Administration*, **50** (7–8), 747–66.

Largely focusing on the production of animated information literacy tutorials using its ANTS (ANimated Tutorial Sharing Project) model, this article describes how a university ensured that its tutorials worked well on a range of platforms, including mobile devices.

Shih, Y. E. and Mills, D. (2007) Setting the New Standard with Mobile Computing in Online Learning, *International Review of Research in Open and Distance Learning*, **8** (2), 1–16.

This paper isn't about libraries, but about mobile learning in general. It discusses the issues in general in addition to outlining a specific case study. The issues and discussions in this paper are transferable to those of us who teach information skills in a library setting.

So, S. (2009) The Development of a SMS-based Teaching and Learning System, *Journal of Educational Technology Development and Exchange*, **2** (1), 113–24.

This paper illustrates a range of learning activities using text messaging, including brainstorming, voting and assessment.

Sutton-Brady, C., Scott, K., Taylor, L., Carabetta, G. and Clark, S. (2009) The Value of Using Short-format Podcasts to Enhance Learning and Teaching. *ALT-J, Research in Learning Technology*, **17** (3), 219–32.

With the use of podcasts being fairly well established, there are many case studies available in the literature. This particular study from the University of Sydney, Australia, focuses on their perceived benefits for learners and shows how these learners prefer to use podcasts. As such, it is a valuable article to start with when considering this technology, as many other case studies focus largely on the technical issues of introducing podcasts, rather than on the end user.

Wang, M., Shen, R., Novak, D. and Pan, X. (2009) The Impact of Mobile Learning on Students' Learning Behaviours and Performance: report from a large blended classroom, *British Journal of Educational Technology*, **40** (4), 673–95.

Often when we talk about using mobile technologies – or even active learning in general – in information skills teaching, the objection is raised that 'my learners wouldn't do that'. This study is therefore included in order to show the difference that using mobile devices in teaching can make to learners in an educational environment that is traditionally passive (a Chinese classroom). This article describes the introduction of a mobile learning system at Shanghai Jiaotong University.

8

E-books for mobiles

Introduction

E-books have been provided through libraries for some time, primarily as texts to read online, through your computer's browser or using a dedicated program. With the advent of electronic ink, e-readers (especially the Kindle), tablet computers and other mobile devices with crisp, clear screens, there has been a shift to users expecting to be able to download e-books from libraries to their own devices. For example, the proportion of US adults who own an e-book reader doubled in six months from November 2010 to May 2011 (Purcell, 2011). This trend continued during the year, with the Kindle, Amazon's flagship e-book reader, selling over a million units a week during December 2011 and becoming Amazon's biggest-selling product in 2011.[1]

Libraries have struggled to meet the demand for downloadable, mobile-friendly e-books because they face many problems in providing e-books in mobile-friendly formats and the whole area is changing fast. Most of the problems can be placed into one or other of two categories: format and licensing. In this chapter we will discuss some of the issues and give examples of the successful provision of e-books for mobile devices.

There are many dedicated e-book readers on the market, some of which have associated applications that can be downloaded to mobile phones and to tablets such as the iPad and that enable you to buy and read books from the supplier of the app, even if you do not own its dedicated device. Some online e-book services are usable on a range of mobile devices. Although dedicated e-book readers such as the Kindle have recently been selling extremely well and some retailers are claiming that e-books are now outselling hardbacks, this chapter will not concentrate on specific handheld readers. Rather than being too specific about devices, formats and platforms, we will focus mostly on generic issues rather than on specific current solutions.

Formats

There are a host of competing formats for e-books, some of which are dedicated to certain devices, others of which are generic and will work across many different technologies (Table 8.1). From a library's point of view, it can be risky to commit to a proprietary format that will work with only one manufacturer's product. However, it can also have benefits, allowing a service to be introduced initially in a focused way, rather than try (and probably fail) to introduce a perfect service that pleases everyone.

Table 8.1 *A comparison of some dedicated e-book readers and the formats they can display*

Device	Formats readable	Apps available
Amazon Kindle	Amazon (.azw); Word Document (.doc); websites (.html); audio (.mp3); Adobe Acrobat (.pdf); text (.txt)	For Apple, Android and Blackberry devices
Barnes and Noble Nook	EPUB (.epub); websites (.html); Adobe Acrobat (.pdf)	For Apple and Android devices
Sony Reader	EPUB (.epub); Word Document (.doc); websites (.html); audio (.mp3 and AAC); Adobe Acrobat (.pdf); text (.txt)	For Android devices

Adobe Acrobat (PDF) files are supported by most e-readers and by most other mobile devices, but they are a far from ideal format for reading on screen. The format preserves the layout of a page, allowing it to be printed out well. However, this means that if you want to zoom in to the text it cannot normally be automatically reformatted to the new page size. This means that delivering e-books (and journal articles) to e-readers in PDF format will work, but is unlikely to be a good experience for the person reading the content.

Perhaps the most effective format for reusable content across many different devices is EPUB. This is a free and open standard that allows functionality such as reflowing (where text flows over the page when you zoom in) and is usable by many e-readers and much e-book software. Although there is no digital rights management (DRM) (to prevent copying of content) built into the standard by default, the standard does allow DRM as an additional layer. DRM support allows publishers to use the EPUB standard without worrying about illegal copying of e-books, though it also makes it considerably harder for libraries to loan e-books.

Licensing

This takes us on to licensing issues. Publishers need to make money in order to survive and are, understandably, concerned about any potential illegal copying of their content. Copying is potentially much easier and cheaper in the electronic environment than it is from physical books, so publishers have tried various ways of making books available electronically while stopping users from copying them.

DRM is one way of doing this. Each copy of an e-book is treated as a unique item and restrictions can be placed on it, including preventing it from being viewed on more than one device and preventing the copying or printing of more than a certain percentage of text from a book. As the popular e-readers often have associated apps for other mobile devices, the DRM protection is increasingly being associated with an account rather than a specific device, allowing people to view their books across all their mobile devices.

The other important way in which publishers have tried to prevent copying of their material is by providing proprietary platforms on which to read their books. Hence, to date, most of the e-books that libraries purchase have been viewable online only, normally through a platform provided by the publisher. Trying to read e-books from standard online platforms using a dedicated e-reader is unlikely to work at all, and on other mobile devices it will be a mixed experience, rarely working well, although this is improving as suppliers introduce more mobile-friendly interfaces.

Publishers struggle to work out ways of licensing libraries to loan e-books to their members, at the same time building in licensing restrictions to ensure that their rights and opportunities to realize future sales are protected. Restricting the number of times that an e-book can be lent has even been suggested, so as to reflect the 'wear and tear' that would result in a popular print book's eventually being replaced (Kingsley, 2011). One restriction of particular concern is to require UK library users to physically go to a library to download an e-book to their mobile device, instantly defeating many of the benefits of an *electronic* book (ebook news, 2010).

Most licensing agreements for e-books purchased for e-readers (i.e. for personal use) assume that the book will be read by only one person, though some now include the ability to 'loan' e-books a limited number of times to friends and family. They do not normally allow libraries to loan books, and separate licensing agreements must be sought. For example, it is not permitted under normal licensing agreements to download popular library books to a number of e-readers and loan the readers out to library patrons.

Case study: Preparing your library for mobile devices
Mandy Callow and Kaye England
University of Southern Queensland Library, Australia

Introduction

This case study had its beginnings in a discussion at the University of Southern Queensland (USQ) Library about the necessity, or not, of providing information on the Library's website about how e-books can or cannot be used on mobile devices, specifically, e-book readers.

In this world of constantly changing technologies, library staff are under continual pressure to adapt to and understand new technologies. Staff are finding it increasingly difficult to keep up with and become proficient users of new technologies. It is safer to assume that your staff do *not* know about mobile devices. As one USQ staff member replied when asked what she had learnt from mobile devices training: 'I learnt that I'm a dinosaur!' Consequently, a small project was created to investigate issues relating to e-books and mobile devices and the best way of communicating this information to Library staff and University staff and students.

E-books in the Library

We quickly became aware that there were two different issues that were confusing staff and students. The first one was mobile devices, what they are and what they can be used for – specifically, within the USQ Library context. The second was the issue of using e-books. It quickly became very clear that the whole matter of e-books is confusing for everyone – dinosaurs or not!

We hear a lot these days about the wonder of e-books and being able to download them onto mobile devices in order to read them anywhere and at any time. In the commercial world this might be so – but it is not the case in the academic world. E-book readers don't seem to be impacting greatly on our services at USQ, though other mobile devices do.

Identifying training needs

In order to provide innovative services and exploit technology, 'we need to understand and play with technology ourselves' (McDonald and Davio, 2011, p. 3). Conducting staff training in order to familiarize staff with various aspects of e-books and mobile devices was vital. Prior to the training, staff were surveyed so as to discover their current knowledge of e-books and mobile devices, what exposure they had had to them, what types of support questions they were getting from students and how confident they felt of being able to deliver support to students.

The results of the survey were not unexpected. Overall there was some confusion and misunderstanding about what e-books were and what their interrelationship was with mobile devices.

Most staff had a reasonable grasp of what a mobile device is. All had heard of the Apple range of iPod, iPhone and iPad. Significantly fewer had heard of e-book readers, especially the Kobo and Sony PRS, and only two actually owned one. Most staff indicated that they had seen students using one of the Apple devices, particularly the iPhone, though as one staff member commented: 'I may have seen them using a device, but I would not have known what it was – just one of those new-fangled gadgets.'

Most respondents had a general knowledge of e-books and had heard of the main e-book aggregator databases to which USQ Library subscribes. The main problems that staff saw with the databases was the confusion caused by the different platforms and the problems with printing, copying and downloading e-books.

Staff were asked if they had been asked specific questions by students and if they would be able to answer them. Fortunately, the most commonly asked questions identified by staff were also those that staff felt they could answer:

- How do I connect wirelessly to my mobile device?
- How do I print this e-book?
- How do I download e-books to my PC?

The most telling response came in answer to the question asking how confident staff felt in answering questions relating to e-books and mobile devices. The majority of staff indicated only 'so-so' as their confidence level, and none indicated that they felt very confident.

What we learnt from the survey

Most staff had a general understanding of what a mobile device is and what it is used for, but knew little of the specifics of using devices or helping students to use mobile devices to connect to USQ networks and use Library resources. Most staff had a general idea of what an e-book is, but were unsure of the specifics of Library-owned e-books and how students use them. We tried to address this in the training offered.

Delivering training

The training was delivered in one 90-minute session. We attempted to incorporate various aspects of e-books and mobile devices. Our main goals for the session were to:

- acquaint staff with the different types of library e-books and give them an understanding of the differences from general e-books
- acquaint staff with new technologies and let them have hands-on experience with mobile devices (specifically iPod, iPhone, iPad, Kobo and Kindle)
- acquaint staff with the USQ mobile interface
- train staff in how to connect mobile devices to the USQ wireless network
- train staff in the basic use of the main e-book aggregator databases
- show staff how to reference different types of e-books.

We wanted this training to equip the staff with sufficient skills and confidence to be able to handle the basics themselves.

How successful was the training?

To ascertain how successful the training was, we asked staff to complete a follow-up survey, based largely on the original survey. Whilst overall there were improvements in staff knowledge and ability, the results were not as good as expected. Staff confidence in answering e-book and mobile device-related questions increased slightly. There were noticeable improvements in staff's perceived ability to answer questions from students, but the results were also not as great as would be expected following training.

Most perplexing were the results for questions asking staff what mobile devices and e-book aggregators they had heard of. During the training, five different mobile devices were shown and passed around the attendees, and basic instruction was carried out in four e-book aggregator databases. In the follow-up survey, some staff who had attended the training indicated that they had not heard of all of the mobile devices and e-book aggregators demonstrated during the training.

As well as not delivering the anticipated results, in some respects the training had actually further confused staff. The discussion that had prompted this training, many months prior to its delivery, had revolved around the issues of accessing Library e-books on mobile devices. During our research we had discovered that there were many other issues about e-books and mobile devices, and thus we had attempted to include all of them in the training. The result was that adequate time and explanation was not allotted to each of these issues during the training. While e-books and mobile devices are intrinsically linked in some ways, in others they are two very different beasts, and this was perhaps not adequately explained to staff.

However, the comments from staff in the follow-up survey were overwhelmingly positive. They showed that staff appreciated the hands-on playing with the technology and that they found the demonstration of different e-book aggregators very useful. Most indicated that they would like further and repeated training in this area.

Lessons learnt

We have learnt a great deal from this initial training and the associated surveys and will be revising our training accordingly, so as to more effectively deliver the necessary information to staff. The main lessons learnt were:

- Do not try to cover too much in one session.
- Create a series of small training sessions delivered over several weeks.
- Deliver separate sessions on e-books and mobile devices, and then a session that explains their interaction.
- With new concepts and technologies, allow staff plenty of time to absorb information.
- Give hands-on training wherever possible.

What we hope to deliver with our new training programme:

- practical, hands-on sessions on the use of different e-book aggregators
- sessions on searching and finding e-book resources
- regular technology expos (every 6–12 months) where new devices and technologies are demonstrated and, wherever possible, staff are able to 'play' with the devices
- practical sessions on resolving issues and problems that students have with mobile devices
- overview sessions on new and emerging Library e-resources.

In the follow-up survey some staff also indicated that they were not asked very often for help with using mobile devices and e-books, and thus would forget what had been covered in training. So we also will be providing:

- cheat sheets at the service desks, detailing basics for different e-book aggregators
- information on the Library website
- short, online tutorials on using different e-book aggregators (for staff and students)
- short tutorials on connecting mobile devices to wireless
- staff loans of mobile devices, so that staff can experiment with using them in different aspects of their everyday work.

The future

Whilst staff may be overwhelmed by the rapid pace of technological change (as one staff member put it, 'Stop the electronic advance, it's snowballing me!'), we cannot afford to ignore new technologies. From this project we have learnt that, in order to provide a usable Library, it is vital to educate both staff and students, particularly in new and emerging technologies and resources. We cannot assume (as our system staff have) that users and staff will become familiar with new technologies as their use in society becomes commonplace.

References
McDonald, S. and Davio, R. (2011) No More Us and Them: mobile support for clients and staff at UTS Library. Paper presented at the CCA-Educause Australasia 2011, Sydney, https://ocs.arcs.org.au/index.php/educause/ccae2011/paper/view/292

Note
This case study is based on a paper presented at the M-Libraries conference in Brisbane, 2011. The full paper can be downloaded from http://eprints.usq.edu.au/19095/.

Ways of providing e-books for mobile devices

As publishers struggle to work out licensing models and systems that they are happy with, together with the formats that can support them, several ways are emerging for libraries to provide e-books for mobile devices.

Services such as Overdrive (http://overdrive.com) or Freading (by Library Ideas – www.libraryideas.com/freading.html) offer e-books, audiobooks and other content to a range of mobile devices, using the EPUB format to ensure compatibility across many different platforms globally. Overdrive deals with a large number of publishers, offering them a means of resolving their licensing and formatting issues in a standard way while providing libraries with a single source from which to buy large amounts of content, rather than negotiate agreements with many suppliers. Overdrive has even recently started to offer electronic books in Kindle format. Overdrive seems to be by far the largest supplier of loanable e-books in mobile-friendly format and a large number of blog posts are available that describe early experiences of using this service; for example Chicago Public Libraries provides a generally positive account of Kindle e-books via Overdrive.[2]

An alternative, large-scale service is Open Library,[3] a project of the Internet Archive (www.archive.org/) available globally, which contains largely out-of-copyright texts, but has additional scanned items from member libraries. At the present time, Open Library also has a reciprocal agreement with Overdrive.

Regional organizations are also developing alternative means of providing e-books for loan, and these may turn into larger-scale offerings in the future. For example, in the USA the Colorado Independent Publishers Association is partnering with a local public library service (Douglas County Libraries) to loan e-books to local library patrons, a service that launched towards the middle of 2011.

It is still early days for services such as Open Library, Overdrive and others. Changes in publishers' agreements can still have sudden and serious effects, such as Penguin's withdrawal from Overdrive (Polanka, 2012) – an agreement that would have been an important part of an offering for many public libraries globally. Penguin is only one example, there is a continual churn of publishers offering and withdrawing content through e-book suppliers, or putting additional restrictions on the use of their titles.

Amazon is also experimenting with loaning books directly to owners of Kindles and Kindle Fires through its Amazon Prime service. At the service's launch, many major publishers declined to participate – a reflection of their general concerns over the impact of e-book lending programmes on sales, much as they have been slow to make e-books available more generally for libraries to purchase.

Publishers seem unsure about what to offer to libraries and have varying restrictions in place. For example, HarperCollins introduced rules limiting libraries to loaning an e-book 26 times before repurchasing it. In 2010 in the UK, the Publishers Association also stated its position as allowing users to download an e-book only from within library premises, rather than online at a distance – instantly destroying a great deal of the convenience of e-books in one fell swoop!

In the current situation of an immature market with limited large-scale options, of publishers unsure what to offer to libraries and often not collaborating, and booksellers setting up e-book lending in competition to libraries (e.g. Amazon), times are confusing for libraries that want to offer e-books to mobile devices. Things are starting to become easier, but for any librarians considering setting up such as service, perhaps the best advice could be to look at their current e-book suppliers, talk to colleagues in a similar position in order to get a flavour of the current issues, and be prepared for everything to have changed in 12 months' time!

Academic e-books

Though services such as Overdrive are, one hopes, starting to simplify the provision of e-books for loan via mobile devices, many academic and school libraries have needed to go down a slightly different route. Sometimes teachers or lecturers want their students to have a course reader containing a selection of books, journal articles and other materials to support their learning. This is often a simple matter when making a print version of such a reader, but not so easy when creating an electronic version.

Creating a course reader in a format suitable for mobile devices can be difficult and time consuming. Even when the materials are available in print form within the library, producing them in e-format is rarely as simple as scanning them and making them available to students to download to their own devices. Copyright clearance must be sought in most countries and great care must be taken to avoid infringing copyright rules. This can often be done for journal articles and book chapters, but tends to require special agreements with the publisher for full books. The market for downloadable full-text books tends at present to be aimed more at individuals than at institutions.

A scheme currently running at the University of Leeds in the UK (Gould, 2010) gave all fourth- and fifth-year medical students iPhones for their studies, partly as a means to access text-books while away from the university. The scheme had to purchase and install apps to provide access to the text-books on these devices, in effect buying a copy of the book for each student.

A wide range of publishers and other services have been experimenting with providing text-books directly to students, the latest being Apple, which is promoting recent changes in its iBooks platform as a way of providing more interactivity than is offered by a 'standard' text-book. It is to be hoped that, as the market develops, these providers will also start to make it easier to buy institutional versions of the same materials that can be distributed to all eligible users.

Large commercial providers of e-book packages to libraries are starting to recognize that library users want to read their e-books on a range of devices. Ebrary, for example, no longer restricts the reading of its e-books to its own on-screen reader and has recently started to permit downloading of its books in a range of formats. If this trend continues it will become considerably easier for libraries to cater for users who wish to read materials on mobile devices. These more established e-book providers still have a variety of models available and the market is by no means mature. It may well be an attractive option for many libraries to work with these established e-book suppliers as they expand the range of formats available, rather than with a supplier who specializes in the mobile market.

Summary

There are many mobile devices that can be used to read e-books. From dedicated e-readers to smartphones, they may each have their own strengths and weaknesses, but overriding them all are legal and copyright issues for libraries. It is becoming easier for libraries to provide materials that can be viewed on mobile devices, whether through dedicated services or through general e-book packages, but there is a long way to go before the market is mature and stable.

While we wait for a more mature market, hands-on training and sharing of experiences can be especially important. The case study from the University of Southern Queensland gives a great example for many of us to follow, of equipping ourselves with the knowledge and skills needed to cope with users who expect to view our books on their own mobile devices.

Notes

1 www.amazon.co.uk/gp/press/home/2011.
2 http://infinitemonkeys.tumblr.com/post/11994755397/kindle-e-books-at-chicago-public-library.
3 http://openlibrary.org/borrow.

References

ebook news (2010) New UK Library eBook Lending Restrictions, *ebook news*, http://ebooknews.co.uk/85/uk-library-ebook-lending-restrictions [accessed 12th March 2012]

Gould, P. (2010) *'Generation Y' Student Doctors Swap Textbooks for iPhones*, www.leeds.ac.uk/news/article/895/generation_y_student_doctors_swap_textbooks_for_iphones [accessed 12th March 2012]

Kingsley, P. (2011) Ebooks on Borrowed Time, *Guardian*, www.guardian.co.uk/books/2011/mar/06/ebooks-on-borrowed-time.

Polanka, S. (2012) Articles of Interest: Penguin, OverDrive, and libraries. *No shelf required*, www.libraries.wright.edu/noshelfrequired/2012/02/09/articles-of-interest-penguin-overdrive-and-libraries/.

Purcell, K. (2011) *E-reader Ownership Doubles in Six Months*, Pew Internet and American Life Project, http://pewresearch.org/pubs/2039/e-reader-ownership-doubles-tablet-adoption-grows-more-slowly .

Further reading

Aaltonen, M., Mannonen, P., Nieminen, S. and Nieminen, M. (2011) Usability and Compatibility of E-book Readers in an Academic Environment: a collaborative study, *IFLA Journal*, **37** (1), 16–27.
A Finnish case study that discusses in particular the problems with licensing materials, as opposed to open access or copyright-free material. It illustrates that often it is not the technology that is the key issue, but

the legal problems with providing useful material on such a device.

Brown, R. (2011) Student Acceptance and Use of E-reader Technology and E-books as an Alternative to Textbooks, *Proceedings of the Academy of Educational Leadership*, **16** (2), 5–9.

Largely a literature review, this pulls together a range of studies that discuss whether or not e-books and e-readers are an acceptable alternative to print materials, in the view of students. As a starting-point, it provides references of many interesting studies.

Clark, D., Goodwin, S., Samuelson, T. and Coker, C. (2008) A Qualitative Assessment of the Kindle E-book Reader: results from initial focus groups, *Performance Measurement and Metrics*, **9** (2), 118–29.

An early study on the usefulness of the Kindle e-book reader in an academic setting. The study gave a number of library users a Kindle, plus $100 to spend on books. Even though the users could buy their own books with the money (rather than worrying about licence agreements to borrow books), the e-readers had a mixed reception. The study concluded that although the devices were received well for reading fiction, there were too many issues for it to be used as an academic tool.

Hamilton, B. (2011) Why We Won't Purchase More Kindles at the Unquiet Library, *The Unquiet Librarian*, http://theunquietlibrarian.wordpress.com/2011/07/27/why-we-wont-purchase-more-kindles-at-the-unquiet-library/.

A fairly detailed blog post about some problems encountered with the introduction of Kindles in Buffy Hamilton's library. This is particularly useful not just for the blog post itself, but also for the long follow-up discussion in the associated comments.

Mallet, E. (2010) A Screen Too Far? Findings from an e-book reader pilot, *Serials*, **23** (2), 140–4.

An account of an interesting pilot study that gave out e-book readers to a range of students and largely left them to do whatever they wished with them, rather than expecting them to use them in a particular way. One of the findings of the study was that PDFs often displayed badly, which has encouraged the university to produce its course materials in EPUB format so that they will display well on such devices.

Polanka, S. (2011) *No Shelf Required: e-books in libraries*, Chicago: ALA Editions

and

Polanka, S. (2012) *No Shelf Required 2: use and management of electronic books*, Chicago: ALA Editions.

These two books bring a wide range of case studies and issues together to discuss how libraries from all sectors can deal with their electronic book collections. They include material on the use of e-books on mobile devices.

Shurtz, S. and von Isenburg, M. (2011) Exploring E-readers to Support Clinical Medical Education: two case studies, *Journal of the Medical Library Association*, **99** (2), 110–17.

There has been a great deal of interest in using mobile devices to support ward rounds in medical settings, as well as medical education. There is traditionally a reliance on a large number of heavy reference texts to support medical students and doctors, making a single, light, easily cleaned device a desirable replacement. This article reports on one study that used Kindles in medical settings.

Tees, T. (2010) Ereaders in Academic Libraries: a literature review, *Australian Library Journal*, **59** (4), 180–6.

This pure literature review pulls together a good assortment of the many trials of e-book readers and gives a good synthesis of the findings.

Walsby, O. and Snelling, M. (2011) E-reader Pilot Projects at the University of Manchester, *SCONUL Focus*, 53, 34–7.

An outline of some e-reader pilot projects in a large UK university library service. In particular, it used some e-book readers to provide course readers for some modules studied at the university. A couple of common problems are highlighted in this study. First, scanned materials do not always display particularly well on these devices, especially if the user wants to zoom into a text. Second, copyright restrictions can make it difficult to provide the full text of materials in an electronic format on such a reader.

So what now?

We've covered a wide range of subjects in this book. We've discussed what mobile services students might want. We've talked a little about how information literacy may vary between 'fixed' and 'mobile' information use. We've shown a range of ways that you could consider of introducing mobile services in your libraries, from text messaging to Augmented Reality, including making your library staff more mobile along the way. But now that you are reaching the end of the book it is time to consider what you should be doing next. Where could you start when thinking about using some of these ideas in your own library?

This final chapter provides an outline for you to use in introducing mobile services into your own library. By the end of this short chapter you should be in a position to decide what steps you should be taking next.

Consider what your users want ... and what your staff can deliver

The worst thing you could do as a result of reading this book would be to pick an idea that seemed exciting to you and immediately try to implement it. Even doing nothing could be preferable, as at least this would not use up any goodwill amongst your staff or users. Instead, the first step should always be to consider your users' needs. Consider how they use your services at present and how they may want to use your services in future. This could be done by looking at existing feedback they have given you, or by running focus groups, surveys or questionnaires. It could even be done by comparing your service to other, similar library services that have implemented mobile-friendly services and seeing how their users reacted. Choose the route that suits you, your library and your users. The key thing is to consider your users' needs before acting.

New mobile services need to be viewed as useful by your users from the start, especially if there is some barrier to their use, such as downloading an app or learning how to use a new tool. By making sure that you know what your users want you can also make sure that any new service will be welcomed and used by them. This will also help you in promoting any new service. If you know why users may desire the service, use this to promote it to the wider population who use your library.

Once you know what services are wanted and seen as useful by your library's users, go for the easiest options first. This might be using text messaging in some way, or making certain information on your web pages more accessible to mobile devices. Taking the easiest – but still desirable – options first will give you quick results and start to create an impression that you are becoming a mobile-friendly library. On the other hand, spending a lot of time and effort to create an Augmented Reality app from scratch might seem a little strange in a library that offered no other mobile-friendly services, and might be a risky option. Failure with early mobile projects would make it considerably harder to continue, whereas success with some easy ones might well pave the way to more ambitious projects in the future.

Mobile services can seem a distraction to some more traditionally minded staff in our libraries, and cheap and easy-to-deliver services can also help to win over these staff. If they can see that there is nothing scary about considering mobile use in the library, and that it needn't take up much time or effort to do so, they will be more likely to be persuaded of the utility of mobile-friendly services. Involve as many of your staff as possible in new services, especially if you expect to encounter resistance to change. Even staff who have never sent a text message in their life can be using a text-a-librarian service within minutes, as long as they are drawn in and trained in a non-intimidating environment.

Consider first services that your staff can deliver easily and understand themselves with a little training. If your staff can't understand a service, then they are unlikely to be successful ambassadors for it to your users.

If you can deliver services that you know your users want and that your staff can readily deliver, it will make an incredible difference to the success of your early mobile projects.

Start steadily ... but don't pilot

For most mobile services, the time for pilots is past. Although technology seems to be developing ever more rapidly, many of the basic possibilities have been available for some time. There are vast numbers of pilots and early pioneers for most of the possible mobile-friendly services. The evidence is available for you to reuse through this book, through many published case studies in journals and blogs and through personal experiences that you may be able to draw on through your professional networks.

Although it is useful to start steadily with services that you know your users want and your staff can readily deliver, start with full services, not pilots. There is no need to pilot text messaging with a small group of patrons: introduce it for everyone. There is no need to give a tablet computer to one member of staff to try roving: give one to all your staff who may be in that situation.

Many other people have already done the necessary pilots to show that these mobile services work. Have the confidence to introduce mobile services fully from the start, rather than trying to replicate the pilots that others have already done. You don't need to prove that these things can be done; you just need to know what your users want, so as to ensure that you are introducing the right services for your library.

Unless you are introducing something truly innovative (which few of us ever manage to do), have the confidence to build on the evidence of the many mobile projects already published.

Nothing is final ... review and assess as you go

If you introduce full services instead of pilots, this doesn't mean that you can forget about gathering evidence on how the services work. As with any service, particularly new ones, you should always seek to tweak and adjust them in order to ensure that they really are what your users want and that they are being delivered in the best way.

Make sure that you build in ways of collecting evidence from your staff as well as from users for any new service. Use this evidence to adjust what you offer and how you do it in the early days, as well as to provide evidence to senior management that any investment in time and money was worthwhile.

Once you are past the early stages of any service, don't forget to carry on reviewing. Nothing is ever final, especially in a mobile environment, where the technology changes quickly. Services that were impractical today may be easy in a matter of months, and those that were important today may become irrelevant just as quickly. Keep reviewing mobile services as you go, as you should do with any service, making sure that you offer only those services that are still relevant and useful to your users.

Keep an eye to the future ... but there is no need to break new ground

With technology constantly changing, don't take the ideas of what is possible and practical today as being fixed. Keep an eye to the future as the mobile environment continues to develop – probably in ways that none of us could reasonably expect. This will help you to keep returning to your users, asking them relevant questions about how they would like to interact with the library and what services they would find it useful to be developed.

This awareness of new technologies and the services that they may enable will help you to prepare for the future – but don't feel that you have to break new ground yourself. It is great if you have the enthusiasm, commitment, time and money to do so, but for most of us there is an alternative. Instead of being in the forefront of mobile innovation, try instead to watch new developments. Read the literature, keep an eye open for case studies and attend training and briefings whenever you can. Take advantage of those people who are in a position to be truly innovative. As long as you keep an eye on the innovators and keep talking to your users, you will be able to pick the new mobile services that look practical for your library and desirable for its users.

Build possible new services into your library's plan if you are in a position to do so, using the experiences of others to select what services are desirable. This will help you to plan the introduction of new services at the time when large numbers of your library's users, rather than just the first early adopters, will be in a position to take advantage of them.

Summary

At the core of every service should be consideration of what its users want, and this is especially important when considering new and different services. So, first of all, consider what your users want. Then you can introduce services (not pilots) that you know are likely to be successful. Continue to assess and develop your services, with the users' needs at the centre of this assessment. Finally, keep a weather-eye on new technologies that are developing and on how early innovators are taking advantage of them.

If you do all of this you will be in a position successfully to introduce and deliver more services in a mobile environment and, importantly, to continue to do so for a long time to come.

Use the ideas and case studies in this book to help you to decide what may work for your library and its users, remembering that these are simply a selection of current possibilities. I hope that they will be useful for you as a source of ideas, suggesting some services that your users will find useful, and that you will continue to watch the multitude of future possibilities as they unfold in front of you.

Index

DATE DUE
